Pipeline Spatial Data Modeling and Pipeline WebGIS

Zhenpei Li

Pipeline Spatial Data Modeling and Pipeline WebGIS

Digital Oil and Gas Pipeline: Research and Practice

Springer

Zhenpei Li
Department of Surveying and Mapping
Engineering
Southwest Petroleum University
Chengdu, Sichuan, China

ISBN 978-3-030-24239-8 ISBN 978-3-030-24240-4 (eBook)
https://doi.org/10.1007/978-3-030-24240-4

This Springer imprint is published by the registered company Springer Nature Switzerland AG
The registered company address is: Gewerbestrasse 11, 6330 Cham, Switzerland

Preface

At present, the construction of long-distance pipelines has shown a trend toward large-scale, systematic, and networked developments. With the continuous expansion of the pipeline construction scale, the traditional pipeline construction management concepts, means and methods are increasingly unable to meet the needs for today's pipeline construction and operation management.

The idea of Digital Oil and Gas Pipeline (hereinafter referred to as, Digital Pipeline) is derived from the concept of Digital Earth. According to the definition of Digital Earth, Digital Pipeline can be defined as "a virtual representation of the pipeline that can collect natural and human information of the pipeline, and enable people to explore and interact with it". The goal of Digital Pipeline construction is to adopt modern, scientific, and digital management of the pipeline design, construction, and operation, with high-tech means throughout the life cycle of the pipeline. The introduction to the concept of the Digital Pipeline provides new means, methods, and ideas for the construction and management of long-distance pipelines.

With the introduction of Digital Pipeline, many pipeline companies and institutions have carried out research on its connotation, construction content, and other aspects, and have worked out specific applications. However, there are still many problems in the current Digital Pipeline construction due to reasons such as technical difficulties and lack of accumulation of historical data, which are mainly reflected in the following ways. First, the idea of Digital Pipeline is mainly applied to the stage of survey and design and construction of the pipeline. It has not been well implemented and applied in the operation management stage. Second, most existing pipeline data models lack modeling for the important pipeline businesses including fire protection, repair and maintenance, pipeline real-time data, automation, or fundamental geographic features. They have only limited support for these pipeline businesses. Third, it is usually developed for stand-alone or local area network application, which is not conducive to sharing pipeline data or expanding the application range of Digital Pipelines. Last but not least, pipeline data is limited to 2D display instead of 3D visual management.

With the rapid development of information and network technology, the author believes that distributed applications oriented to the network will be one of the main features of the Digital Pipeline applications and that the network Digital Pipeline will be the development direction of Digital Pipeline construction. Based on this point of view, combined with the problems existing in the current Digital Pipeline construction, the author proposes the concept of "Web-based Digital Pipeline". The Web-based Digital Pipeline focuses on the operation management of the pipeline. Its core idea is to combine computer network, Web Geographic Information System (WebGIS), GIS Web Services, pipeline Supervisory Control and Data Acquisition (SCADA), OLE for Process Control (OPC), network virtual reality, and other advanced systems and technologies in order to realize Web-based release, query, and management and analysis of pipeline information. It will also realize remote, multilevel distributed monitoring, Web-based 3D visualization, and virtual reality representation of pipelines.

According to the construction objectives of Web-based Digital Pipeline, the author has carried out research on the implementation and application of the Web-based digital pipeline system. The research and application results are described in the "Digital Oil and Gas Pipeline: Research and Practice" book series. The main contents of this book series are as follows.

(1) Establishment of Pipeline Spatial Data Model (PSDM).
By carrying out the demand analysis of pipeline and its surrounded data and using the object-oriented methodology, Pipeline Spatial Data Model (PSDM) is established based on ArcGIS Pipeline Data Model (APDM), as well as the design experience of other present main pipeline data models such as Pipeline Open Data Standard (PODS) and Integrated Spatial Analysis Techniques (ISAT).
In the digital pipeline system, the pipeline spatial database is the core part. The key to build a pipeline spatial database is to design a good Pipeline Spatial Data Model. The pipeline data model formulates the basic data structure and behavioral characteristics of the pipeline data. It not only relates to the behaviors and events that occur in the pipeline construction and management process but also involves the situation around the pipeline. The Pipeline Spatial Data Model fully considers the spatial distribution characteristics of pipeline data. It models the attributes and behaviors of related data along the pipeline, as well as the relationship between pipeline spatial data and attribute data. The Pipeline Spatial Data Model also defines rules for storing spatial data in a business relational database, enabling the Pipeline Spatial Data Model to fully utilize the powerful management functions of the business relational database. Research is conducted on support and implementation methods of the Pipeline Spatial Data Model for the pipeline real-time parameter data, the linear referencing system, and dynamic segmentation technology. The object-oriented design ideas and methods are adopted. The PSDM is designed using the object-oriented methodology, so as to promote the reusability and extensibility of the model. PSDM adopts module designs for pipeline elements. Several

important modules such as automation, fire protection, repair and maintenance, and fundamental geographic elements are added to PSDM.

(2) Research on the implementation methods of pipeline WebGIS system.
The pipeline GIS system belongs to an applied geographic information system. The main development methods of applied GIS are analyzed and compared. The compared conclusion is that the Component GIS-based development method is suitable for the development of the digital pipeline geographic information system. The application of Component GIS in network environment is also studied. The author summarizes the implementation methods and limitations of traditional WebGIS, and proposes a WebGIS implementation method based on Web Services and Component GIS. Web Services is used as the application framework to publish GIS functions. It is implemented by Component GIS, and then the GIS function published by Web Services, together with ArcGIS Server, is used to build the pipeline WebGIS system. This method can not only realize the GIS interoperability by using Web Services but also has the advantages of Component GIS, such as flexible structure, low development costs, high performance, and reusability.

(3) Research on integration method of pipeline SCADA system and pipeline GIS.
Based on the analysis and comparison of the main methods of current SCADA system and GIS integration, the OPC-based pipeline SCADA system and GIS integration method are proposed. A data access component is developed with OPC interfaces to implement the real-time data accessing to the SCADA system, and the real-time data transfer to PSDM. In this way, the SCADA system provides real-time data of the pipeline to the GIS system through the OPC-based data access component. The GIS system sends instructions to the SCADA system through the data access component. The historical data of the SCADA system is obtained by accessing the historical database of the SCADA system through Open Database Connectivity (ODBC). By doing this, the real-time monitoring of pipelines based on GIS system can be realized. Moreover, combined with the real-time data and historical data of the pipeline SCADA system, relying on the powerful spatial analysis capability of the GIS system, the pipeline operation conditions online or offline analysis or simulation can be performed to provide diversified decision-making support for efficient pipeline management.

(4) Research on the implementation method of pipeline network virtual reality system.
The Pipeline network virtual reality system is an important part of the digital pipeline construction. Its main purpose is to build a network-based and interactive 3D dynamic virtual pipeline to realize network 3D visualization and virtual reality representation of the pipelines. The main research contents of this part include large-scene roaming of pipelines, virtual facility modeling, and 3D visual monitoring. Research is conducted on 3D terrain modeling, terrain model texture mapping, network virtual reality geographic information system construction schemes, methods for improving performance and speed of

large-scene 3D terrain browsing in network environment, interaction methods of virtual scenes and external programs, pipeline 3D visual monitoring through interaction between virtual facilities and pipeline SCADA system, etc. At the same time, the methods of the interaction between the pipeline network virtual reality system and the pipeline WebGIS system at the data level and the UI level are also investigated.

The title of volume 1 of the "Digital Oil and Gas Pipeline: Research and Practice" book series is, "Pipeline Spatial Data Modeling and Pipeline WebGIS". The title of volume 2 is, "Pipeline Real-time Data Integration and Pipeline Network Virtual Reality System". This "Digital Oil and Gas Pipeline: Research and Practice" book series introduces the author's latest research and practice on digital pipeline construction. The series covers the latest research results and technologies in WebGIS, GIS Web Services, pipeline SCADA, OLE for Process Control, X3D, and network virtual reality. The research includes such core contents of digital pipeline construction as the Pipeline Spatial Data Model, the pipeline WebGIS system implementation method, the pipeline SCADA system and GIS system integration method, and the pipeline network virtual reality system implementation method. This book series will be a useful reference for researchers and practitioners engaged in oil and gas storage and transportation, pipeline automation, geographic information systems, virtual reality, and other aspects.

Chengdu, China Zhenpei Li

Acknowledgements

The author would like to thank the following people: Sasha Fan for her translation work for this book, Yang Liu for participating in the preparation work and amending part of sections of this book, and postgraduate student, Lehao Yang, for collating work for the references and contents of this book. A large amount of literature was referred to, some of which had unnamed authors. The author of this book is grateful to all of them for their contribution. Emily Sarah J. Villanueva heavily involved in improving writing of this book.

Contents

Chapter 1
Introduction

Abstract In this chapter, the author introduces the background of emergence of Digital Oil and Gas Pipeline (hereinafter referred to as, Digital Pipeline), the connection between Digital Pipeline and Digital Earth, and the concepts, functions, and significance of Digital Pipeline. The key technologies, business systems, and construction contents of Digital Pipeline are described. The research and application status and current deficiencies of Digital Pipeline construction are also discussed. The author puts forward the concept of "Web-based Digital Pipeline" considering the trend of Digital Pipeline development toward network. The core ideas, construction objectives, and technical architectures of Web-based Digital Pipeline are elaborated in detail. Finally, the main problems and research contents of this book are introduced.

Keywords Digital Pipeline · Digital Earth · Key technologies · Business systems · Construction contents · Application status · Deficiencies · WEB-based Digital Pipeline

1.1 Digital Pipeline: The Emergence of a New Technology

With the rapid development of long-distance oil and gas pipeline construction, it is increasingly difficult for traditional concepts and methods of pipeline construction and operation management to meet the needs of modern pipelines in environmental protection and safety management. Shortcomings do exist in the feasibility study survey and design, and construction and operation management of traditional pipeline construction. Their shortcomings are reflected in the following aspects [1, 2]:

(1) At the survey and design stage, most of 1:50,000 and 1:100,000 topographic maps used are products from the 1970s or 1980s, (there are even some from the 1950s and 1960s). They are inconsistent with the current situation, especially in developed areas. Some geological data are quite outdated and can hardly reflect the most current situations in geological hazard analysis and interpretation, river evolution, mountain change, earthquake rupture, etc., leading to pipeline routes being chosen blindly in an almost simpleminded way. Those immature schemes make it difficult to optimize the determined route. The fundamental data for

Z. Li, *Pipeline Spatial Data Modeling and Pipeline WebGIS*,
https://doi.org/10.1007/978-3-030-24240-4_1

design are also incomplete, risking the quality of the pipeline design. Furthermore, the traditional pipeline surveys, mainly carried out in field, can hardly reach the requirements in the pipeline construction process and the demand for efficient data acquisition and update due to its long timescale, high cost, and low efficiency in measurement, survey, route selection, and location. In addition, the rapid development of the national economy promotes a large amount of infrastructure construction, which results in numerous pipeline rerouting. Meanwhile, as industrial technology improves, the auxiliary equipment, station process flow, and automation communication of the pipeline are being updated frequently, which challenges the drawing updating and modification. In such a sense, those added drawings, mostly with repeated content, increase the workloads in file management. Varied pipeline information is all marked on the paper drawings and needs to be handled with care, otherwise, they may be broken or lost. Besides, the data cannot be employed to the full play because of their poor performance in exchangeability and versatility.

(2) At the construction management stage, the small-scale drawings cannot reflect the actual situation and the design concepts cannot be fully expressed. This leads to different understandings during the construction process, and quality problems may lie in those built pipelines. The design documents for bidding and construction may lack detailed content, resulting in under preparation. Additionally, changes in circumstances ask for corresponding adaptations in design and construction. If the staff cannot communicate with various data in an effective way, the construction progress, quality, and duration may be influenced accordingly. In the construction survey of completion drawings, unreasonable data collection may result in greater inconvenience to the operation management.

(3) At the operation management stage, difficulties may arise in normal operation, routine maintenance, and management of the built pipeline (due to the simple design), changes in construction process, insufficient completion of data, and especially key construction parts not being effectively tracked and regularly inspected. For these reasons, it is hard to provide up-to-date and accurate detailed information in a timely manner and respond to emergencies.

In general, the technical means, work processes, and accumulated documents employed in the traditional pipeline construction, including the survey, design, construction, and operation, cannot meet the needs of today's pipeline construction and operation management. As the pipeline construction scale continues to expand, it is urgent to find new management concepts and methods to upgrade the level of pipeline construction. Under such circumstances, the idea of Digital Pipeline has emerged, thanks to the inspiring concept of Digital Earth.

1.2 The Connection Between Digital Pipeline and Digital Earth

The concept of Digital Pipeline is closely related to the advance of Digital Earth: the ideas and key technology of Digital Pipeline derive from Digital Earth. To some degree, the Digital Pipeline can be reckoned as a specific application as well as an important component of Digital Earth. Therefore, it is necessary to give a brief introduction to Digital Earth.

1.2.1 Digital Earth

Currently, as human civilization has highly developed, people are fairly capable of acquiring information, and we step into the information age with explosive knowledge. For instance, the American National Aeronautics and Space Administration (NASA) Planet Earth Plan brings as much as 1000 GB information every day. The land satellite Landsat obtains a set of global satellite image data every two weeks and has already collected satellite data for over 20 years. On the one hand, humans urgently demand information, while on the other hand, this ready-available data has not been fully utilized. The major problem now lies in understanding the content of this data and applying the information.

At the same time, globalization has become an inevitable trend in the development of global society now. Conducting global research has also become a major task of today's scientific research with the premise that information is shared among different regions and organizations around the world. An existing problem is that various data are scattered in different areas and institutions. Without a uniform standard, data sharing and employment could run into obstacles.

Therefore, in the face of the contradiction of "information explosion" and the difficulty in sufficient use, as well as that of the urgent needs of global information and the complexity of space information, how to establish a mechanism to realize the sharing and efficiently use information resources have become the key subject in global information research. It is under such circumstances that the Digital Earth comes into being.

On January 31, 1998, Al Gore, the former president of the United States of America, officially put forward the concept of Digital Earth in his speech "*The Digital Earth: Understanding our planet in the 21st Century*" [3]. He proposed that the Digital Earth is a "*3D expression of the real earth, which can implant a huge number of multiresolution geographical data*". Since then, the concept of Digital Earth has been rapidly and widely recognized and has gained active responses from various countries and regions, as well as an increasing number of related research [4–8]. However, the concept of "Digital Earth" was not clearly defined.

In 1999, in a global seminar held at the University of Maryland in America, majority of the scholars agreed to define the Digital Earth as "*the virtual expression of the earth produced through collecting the natural and human information around the globe. People will be able to explore and interact with it*".

Li [9] saw the Digital Earth as a unified and digitalized representation and recognition of the real earth and its related phenomena. The core concepts were to use digitalized methods to deal with problems concerning the natural and social activities around the world, to maximally utilize resources and to enable people to obtain information about the earth. Digital Earth is characterized by implanting large quantity of geographical data, and can also achieve multiresolution and 3D descriptions of the earth referred to as the "virtual earth". He supposed that Digital Earth was based on computer technology, multimedia technology, and mass storage technology. With broadband network as the link, it applies mass earth information to describe the globe in a multiresolution method, on a multi-scale and multi-dimensional level. He believed Digital Earth could be used as a useful tool to support human activities and to improve quality of life.

Cheng et al. [10], argued that Digital Earth referred to the digitalization of the earth, and more precisely, the informatization earth, which is consistent with the concept of the national informatization. Informatization is the whole process of digitalization, networking, intelligence, and visualization with computer as the core. To be specific, Digital Earth stands for a kind of technical system which takes the earth as the object. It is based on geographical coordinates integrated with multiresolution, massive, and multiple data fusion, and is represented in multi-dimensional ways (both stereoscopic and dynamic) through multimedia and virtual technology as well as with spatial, digital, networked, intelligent, and visualized features. Digital Earth represents a technical system aiming to realize the digitalization or informatization of the earth, and it can also be interpreted as the digitalized virtual earth. More precisely, Digital Earth refers to a technical system managed by a computer network after the digitalization of the entire earth.

In summary, the primary concepts of Digital Earth can be described in the following three aspects [10, 11]:

(1) Digital Earth refers to the digitalized and three-dimensional display of virtual earth, or an informationized earth, which includes digital, networked, intelligent, and visualized earth technical systems.
(2) The implementation of Digital Earth plan requires the cooperation of governments, enterprises, and academia. It is also a social activity and needs the concern and support of the whole society.
(3) Digital Earth is a new technological revolution. It will change social production and lifestyle, and bring about progress in scientific and technological development as well as in the social economy.

To summarize, Digital Earth is a completely informationized virtual earth. Based on the supporting technologies like the information accession, storage, transmission, expression and processing technologies, it can process mass information of the earth and its related natural and social phenomena according to their geographical coordinates. In this way, people can understand the macro and micro conditions of each corner of the earth quickly, completely and vividly, as well as take advantage of the information to solve various problems of natural and social activities [12–14].

Since the proposal of National Information Infrastructure (NII) and National Spatial Data Infrastructure (NSDI), Digital Earth, acting as the commanding height of the present technology, is a far-reaching technological strategy [10]. It takes earth as a research object and develops it into a high-tech system which makes use of various technologies, especially the information technology to serve people. Digital Earth is regarded as a fundamental technology project in the twenty-first century [15]. A large number of modem technologies are needed to support the implementation of Digital Earth. Gore stated in his report that the core technologies of Digital Earth include the scientific calculation, massive data storage, broadband network, satellite data accessing, interoperability, and metadata.

The proposal of Digital Earth is strategically meaningful. It will bring social and economic benefits to various fields including agriculture, industry, forestry, water conservancy, transportation, mining, communications, education, resources, environment, population, military, and urban construction. As it covers almost every aspect of human life, its potential application is enormous. At present, Digital Earth has been practically applied in global change and sustainable social development, geoscience, urban construction, fine agriculture, forestry, traffic information management, and other fields [16–25].

1.2.2 Application Levels of Digital Earth and the Digital Pipeline

The construction of Digital Earth is a huge information engineering project as it deals with super-mass data. Therefore, it cannot be fulfilled by any government, organization, enterprise, or research institute alone. Instead, its gradual perfection requires cooperation among thousands of hardworking people, enterprises, universities, research institutes, governments of all levels, and organizations, by utilizing available technologies and resources [26].

Digital Earth is a huge system and there is no fast way to establish it. Its construction and improvement need gradual efforts from different levels including nations, regions, and the globe [27, 28]. With continuous research on Digital Earth, its application has covered levels from regional to national and even global, involving fields such as the digital cities, digital universities, digital enterprises, digital communities, and Digital Oil fields [29, 30]. All of these ideas originate from Digital Earth which aims to digitalize the real world by using various technologies like the data accessing, storage, visualization, and information processing technologies, and then, use this digitalized system to solve kinds of practical problems and to improve the information management level of the world. Therefore, Digital Pipeline is an exact application of Digital Earth to the oil and gas pipeline field.

1.3 Digital Pipeline

1.3.1 Concept of Digital Pipeline

The concept of Digital Pipeline derives from that of Digital Earth and is the application of Digital Earth to oil and gas pipeline design, construction, operation and management. Currently, Digital Pipeline has not yet been clearly defined, but according to the common definition of Digital Earth, one possible definition for Digital Pipeline can be "*...the virtual expression of the pipeline, which collects the natural and social information concerning the pipeline. People can explore and interact with it*". This definition contains three important aspects. First, Digital Pipeline is the virtual expression of the real one, and is mainly presented by a computer information system so far. Second, Digital Pipeline not only covers the information of the pipeline itself but also contains the related information concerning it. Third, Digital Pipeline has interactivity. It enables people to supervise and control the acts of the real pipeline and Digital Pipeline can also provide real-time feedback information to people.

Wang et al. [1] believed that Digital Pipeline was based on the information infrastructure and was supported by multi-scale, multi-aspect geospatial information. In line with the concept of Digital Earth, Digital Pipeline integrates pipeline facilities, circumstances, geographical conditions, economic, social, and cultural information within a 3D geographical location via computer, modern surveying and mapping, modern network, virtual reality, and digital communication technologies. It functions as a digital management and decision-making supporting system which runs with great efficiency in data collection and processing in pipeline research, prospective design, construction, and operation management.

The construction of Digital Pipeline aims to realize modern, scientific, and fully digitalized management of the pipelines by adopting high technologies to the design, construction and operation of the pipelines in their whole life cycle. The Digital Pipeline collects a large amount of multiresolution, multi-scale, and 3D geospatial information of the pipelines and their surroundings. It then puts this data into an advanced decision-making system based on geographical mathematics models so as to integrate the information of pipeline resources, environment, society, and economy into an application system. In this way, technicians and executives can be provided with visualized digital service. Digital Pipeline is fully digitalized with a complete pipeline information model. This complete and digitalized pipeline rearranges the information related to the pipeline according to its geographical location so that people can get quick access to complete pipeline information in a visualized manner. At this point, a bank of information is at your fingertips.

1.3.2 Functions and Significance of Digital Pipeline

The functions of Digital Pipeline are mainly displayed in the following aspects [1]:

(1) Digital Pipeline helps to select the pipeline route at the feasibility research and preliminary design stage. It can help construct a 3D image of the pipeline by using the digital image DOM, the Digital Elevation Model (DEM) and other materials. It offers a convenient solution for decision-makers deciding on the most economical and reasonable pipeline routes as well as confirms the most suitable place to lay the pipelines, stations, and valve chambers on the basis of abundant information of culture, geology, transportation, water system, and vegetation offered by the remote-sensing image and aerial survey data.
(2) At the construction stage, the construction units can make use of various information and resources to deploy staff and facilities rationally, to transport the pipelines, and to arrange the labors scientifically. At this stage, information including construction progress, the locations of each crater, and the geographical changes after the backfill of the pipeline need to be added gradually.
(3) At the operation stage, the instant status of pipeline can be dynamically tracked through the information gradually gained at the design and construction stage. Once any breakdown occurs, the Digital Pipeline can help position the breakdown and circle out the influenced areas through fuzzy query and network analysis. It can timely figure out a maintenance plan to ensure the smooth operation of the pipeline.

Digital Pipeline plays a very active role in improving the overall oil and gas pipeline design level, and in updating the construction design, operation, and management concepts so as to construct highly efficient and highly qualified projects. The construction of the Digital Pipeline will be a revolution compared to the traditional pipeline construction methods, and it will also trigger a profound change for the construction concept of the modern pipeline. The gradual application of Digital Pipeline technology will optimize the pipeline route selection, deploy the construction more reasonably and efficiently, and improve the pipeline operation with greater efficiency and better safety so as to promote pipeline exploration and design. Digital pipeline will lead to a scientific and standard construction and operation management system [31, 32].

1.3.3 Core Technologies of Digital Pipeline Construction

The core technologies of Digital Pipeline construction mainly include Remote Sensing (RS), Global Positioning System (GPS), Geographic Information System (GIS), Supervisory Control and Data Acquisition (SCADA), network and multimedia technology, modern communication, and virtual reality technology. RS, GPS, GIS and SCADA altogether are known as 4S technologies [33, 34].

(1) Remote sensing is a comprehensive reflection of the earth ground information and can vividly reflect the landscapes of the earth, geomorphy, especially the latest information about the changes of river, lakes, farmlands, the expansion of residential areas, and other dynamic changes of earth. Besides, the remote-sensing images contain both general and 3D information far beyond any other survey methods. Moreover, its receiving system is stable enough to prevent itself from factors like transportation, weather, terrain, and region, which can help to achieve information in a more efficient way. Today, the remote-sensing technology gets to maturity with lower costs and higher resolution ratio of images. For example, the US Quick Bird has the resolution up to 0.61 m and the dimension of each image can be up to 16.5×16.5 km. The resolution of SPOT-5 reaches 2.5 m. The introduction of remote sensing into the pipeline engineering filed [35, 36] can bring about revolutionary changes to the traditional pipeline survey and design:

- Highly efficient and timely design graphs with various styles, specifications, and scales
- Offering electronic data
- Systematic errors and accidental ones of survey mapping can be greatly reduced
- Design quality and efficiency can be further improved.

(2) The Global Positioning System (GPS) can provide timing and positioning service for the objects of the study. GPS can provide accurate 3D positioning, velocity, and one-dimensional time information at any time around the globe through its navigation, positioning, and timing functions. It has sound competence in anti-interference and in confidentiality and not only offers timing and positioning service for the pipeline construction but ensures quantitative accuracy of the remote-sensing results [37].

(3) The Geographic Information System (GIS) technology, [38] supported by computer software and hardware, is a computer system established specifically for serving geographical research and geographical decision-making. It collects, edits, analyzes, manages, simulates, and displays spatial data so as to offer various, dynamic, and real-time geographical information. Through establishing the geographic information system of the Digital Pipeline [39–41], the spatial positioning, searching and analysis of the pipeline and its facilities information can be achieved. Relying on the detailed pipeline information database (including the basic conditions of pipelines, the pipeline route graph, profile graph, the natural conditions along the pipeline, the traffic situations, the crossing projects, and the facilities of the pipeline), the pipeline tests, and pipeline engineering, the routine inspection of the pipelines can be managed so as to improve and guarantee the operation safety. The geographic information system of the Digital Pipeline also has additional functions like inputting and editing pipeline data, inquiring and searching, data statistics, and analysis of vertical and horizontal graphs.

(4) Developed from computer science, telecommunications, and control technologies, SCADA is the basis of pipeline automation. Its aim is to realize the automatic collection and control of the operational technological parameters of the pipelines and pipeline stations, and to achieve interlocking of furnaces and pumps, controlling of the surge protection, and ensuring the secure operation. The technological parameters needed by Digital Pipeline include the entering and leaving pressure, temperature, flow, water content of the crude oil, and density of the pipeline station; the entering and leaving pressure, temperature, and valve state of the pumps; the electronic parameters of the pump motors (include the three-phase current, three-phase voltage, electric quantity, active power, reactive power). By collecting these parameters, the system can give out two kinds of real-time information: the pipeline integrity status and the economic operation of oil pipeline such as the oil transmission consumption and efficiency.
SCADA system can achieve the full automatic control of the pipeline. Through collecting data by the communications among the Remote Terminal Unit (RTUs), Programmable Logic Controller (PLCs), and other input/output devices based on the hosts and microprocessors, the whole pipeline is under supervision and monitoring which ensures the safe operation and optimized control of the system [42, 43].
(5) Virtual reality technology [44] organizes and manages the pipeline information in order to vividly present it through virtual reality technology. This is one of the main functions of the Digital Pipeline. The virtual reality system uses multimedia and computer technology to create a vivid 3D space or even 4D temporal and spatial sensing space to enable people to use visual sense, auditory sense, smell, tactile, and other sensory organs to implement the real-time participation and interaction with the virtual space. The virtual reality technology can be employed to produce the virtual environment of the pipeline to provide a visualized information space for the users. Through various sensory devices, users can "participate" in the virtual environment, observe, and obtain information of the virtual pipeline space so as to realize the direct and natural interaction between the users and virtual pipeline [45].

1.3.4 Digital Pipeline Business Systems

Digital Pipeline consists of three business systems: survey and design system, construction management system, and operation management system [33]. They are interconnected data sources for each other. The pipeline information database, the core of the whole system, is built to manage, store, and share all the data collected during the whole life cycle of the pipeline.

The pipeline survey and design system is made up of a series of subsystems in geological exploration, route design, cathodic protection, process design, civil engineering design, and general layout design, etc. While the information produced by these subsystems include the society, geography, meteorology, hydrology, geology and survey, remote sensing, professional 2D and 3D design drawings, documents, equipment, and material, altogether they constitute the design information database.

The pipeline construction project management system involves design, construction, material, schedule, expense, document, Quality, Health, Safety, Environmental (QHSE), and other related content. During the process of pipeline construction, the Digital Pipeline system plays a more important role in scheduling, logistics, and cost management.

Pipeline operation and management system is a mixed system composed of multiple subsystems including the completion systems produced at the first two stages, the SCADA subsystem, sales subsystem, and the pipeline resources management subsystem. One of the main tasks of the Digital Pipeline system is to exchange and share the data of these subsystems while protecting their information security.

1.3.5 Construction of Digital Pipeline

1.3.5.1 At the Survey and Design Stage

The major task of this stage is to select proper pipeline routes and to obtain the 4D data within 200 m along the pipelines routes via the remote-sensing and the digital photogrammetry technology. Meanwhile, the GIS and GPS technology is applied to construct preliminary information management systems concerning the topography, circumstances, population, economy, and other information along the pipeline route. The specific construction includes [46]

(1) The remote-sensing image processing system: This system can process, analyze, and display multispectral, hyper-spectral, and radar data of the satellite remote sensing. The interpretation of satellite remote-sensing data provides reliable evidence in the natural environment, geography, geology, and other regional information for the decision-making of the pipeline route selection.
(2) The digital photogrammetry processing system: This system offers fundamental data for pipeline routes selection. Compared with satellite remote sensing, aerial survey data has advantages like larger scale, higher resolution, and more precise details, which play a critical role in route selection.
(3) Digital Pipeline feasibility study system: This system first integrates the data of interpreted remote-sensing image, the digital photogrammetric, the population, environment, economy, and other geographic features, and then estimates the economic benefits of the project and makes the general route plan based on the integrated data analysis.
(4) Geological survey information system: This system provides the ready drawings of the geological, surveying, and hydrological information, and attributes data along the pipeline for pipeline route selection.

(5) Pipeline design CAD system: Since the pipeline route is confirmed, it serves construction drawing designs for the pipeline and its affiliated facilities such as distribution station and valves.
(6) Communication design system: This system lays out the communication facilities of SCADA system.

1.3.5.2 At the Project Construction Stage

The Digital Pipeline can provide a wide range of Internet information service at this stage. For example, the pipeline builders can check the pipeline information and its surroundings along the routes via Internet, as well as check the project progress on a specific day or at a specific procedure. Digital Pipeline construction includes [46]

(1) GPS data acquisition system: This collects geo-coordinate data of the pipelines through GPS during construction.
(2) Survey management information system: This collects and calculates survey data in the construction stage and makes drawings and reports accordingly.
(3) Construction management system: This collects, generates, audits, reports, and manages specific construction data, permanent statistics, and other documents.

1.3.5.3 At the Operation and Management Stage

At this stage, the construction of Digital Pipeline includes [46]

(1) Pipeline operation system: It is an integrated operation and management system of the enterprises, facilities, customers, and commodities. Its operation and management include manufacturing, enterprise administration, public relations, upstream business, marketing, and sales. It connects the Digital Pipeline with its potential markets and customers directly, and best reflects the ultimate economic benefits of the pipelines.
(2) Pipeline facility management system: It integrates the related data of each pipeline construction stage and establishes the pipeline equipment database so as to achieve its digitalized management.
(3) Pipeline integrity management system: It aims to ensure the safe operation of pipelines by applying various integrity management technologies like the pipeline risk assessment, pipeline corrosion, defects assessment and control, geological disaster monitoring, supervision and control, in-line inspection and leak detection, crossing detection and evaluation, and pipeline maintenance.
(4) Pipeline update and maintenance system. The establishment of this system is the necessary premise of achieving pipeline integrity management. It guarantees the pipeline to work safely and efficiently during its life cycle by updating and maintaining various pipeline-related data timely.

1.4 Application Status of Digital Pipeline in the Pipeline Industry

Digital Pipeline technologies were first proposed in the United States in the 1990s and have been successfully applied in some countries. Currently, the technologies are comparatively mature in the United States, Canada, and Italy, where Digital Pipeline technologies such as high-resolution satellite remote-sensing image, digital photogrammetry, geographic information system, and 3D design have been widely used in pipeline feasibility study, survey and design, construction, and operation management. These technologies have played an important role in determining the optimal routes, resource configuration, disaster monitoring and early warning, and operational risk management. The U.S. Pipeline Safety Administration has established the National Pipeline Mapping System (NPMS), which includes data of nearly 190×104 km pipeline in North America, allowing online distribution and sales of pipeline graphics and other related data. Some examples include Buckeye Partners, Enron Transportation Services, SNAM, and Alliance Pipeline. Buckeye Partners has established a pipeline management system based on GIS. Enron Transportation Services has set a pipeline emergency response system based on GIS. SNAM (Italy) developed the Shared and Integrated Geographic Information System (SIGIS) by using GIS technology to realize the collection and management of pipeline information and production of maps of natural gas pipelines throughout the country. Alliance Pipeline also utilizes a pipeline geographic information system providing users with pipeline GIS data [1].

Gasunie, a Dutch Company, adopted Intergraph technology and integrates GIS, enterprise resource planning system, and risk management system, etc., to build an advanced Digital Pipeline system. With professional software, Ruhrgas AG's Digital Pipeline management system shared data across engineering systems and operation management [47].

General Electric and Accenture jointly launched the world's first "Intelligent pipeline solutions" in 2014; Columbia Pipeline Partners LP, one of the largest pipeline operators in the US, applied it in 2015. "Intelligent pipeline solutions" integrates geographic information system, production system, risk management system, and equipment condition monitoring system, as well as external data such as weather, earthquakes, and third-party activity information. The integrated data has an open data structure that provides real-time digital references to the pipeline. With this solution, users can integrate forecasting function to support future pipeline decisions [48].

With the introduction of Digital Pipeline, related institutes in China have carried out research on its connotation and construction content. Ji-Ning tie-line of the West-to-East Gas Pipeline was the first project in China to apply the digital technology for long-distance pipeline [32]. The project used satellite remote-sensing and digital photogrammetry technologies during the survey and design to select the route and to obtain 4D regional data within 200 m of both sides of the pipeline. It also adopted 4S technology and high-tech means, such as computer and multimedia technology, to initially build a Digital Pipeline information system. This includes pipeline facilities, environment, and geological conditions along the route, and economy, social, and cultural background so as to realize the design under visual conditions. The Digital Pipeline information system will provide a network platform combining data and map for pipeline construction and maintenance management. On this platform, pipeline builders can view the visual information of different scales of pipelines and their surroundings via Internet during the construction stage. Pipeline builders can also view the progress of a certain process on a certain day. Every steel pipe in the Ji-Ning tie-line is well recorded with complete data. The data can be traced at each step from first, a semifinished product in the steel plant; second, a finished one in the steel pipe factory; third, the transit transport; and finally, the arrival at the construction site. The basic data related to welding processes, X-ray film files, coordinate values, and depth of welds are also available. All of the data is stored in the pipeline information system, and the source can be detected in case of any problems.

Some literature [1, 2, 32, 33, 49–51] introduced concepts related to Digital Pipeline. Li Shanshan discussed the standardization construction of Digital Pipeline technology system [52]. Yu [53] applied Digital Pipeline technology to pipeline survey, design, and construction management, and Han [54] studied the application of Digital Pipeline technology, especially GIS technology, in pipeline design management. At the pipeline construction stage, Yang Mao et al. provided complete and reliable fundamental data for Digital Pipeline construction. The data included construction process information, construction materials information, and construction documents generated by the pipeline project department, project supervision, material procurement unit, construction unit, testing unit, and other departments through the data acquisition platform of GIS [55].

Tang Jiangang collected the complete measurement data for the Digital Pipeline under construction with RTK measurement technology and controlled the quality of the measurement results with multiple calibration measurements [56].

Dong Rongguo built the integrated eastern China refined oil Digital Pipeline system with pipeline integrity data management, GIS business application, pipeline intelligent inspection, emergency analysis decision, and station 3D visualization, by using two pieces of international leading software, ArcGIS and skyline, and 3S (RS, GPS, GIS) technology [57].

China National Petroleum Corporation(CNPC)used real-time data acquisition and pipe network operation monitoring in projects such as the second West-to-East gas pipeline and China–Myanmar oil and gas pipeline, to achieve centralized monitoring and operation scheduling. Sinopec Group launched its Digital Pipeline construction in 2007 and applied it first to Yu-Ji Pipeline, then extended it on the entire Sichuan-East Gas Pipeline in which a professional pipeline GIS was built with the data covering all pipelines and ancillary facilities [58]. China Petroleum Pipeline Engineering Corporation used digital design means at the construction drawing stage of the Ha-Shen line, and for the first time, handed over 71 data forms (including traditional design results) of 16 fields in the feasibility study and preliminary design stage to the pipeline life cycle management database. China Petroleum Pipeline Engineering Corporation also carried out the digital recovery of in-service pipelines, taking the lead in digital handover and digital recovery of in-service pipelines. In 2016, the corporation completed the digital design of Thailand's No. 4 line compressor station, the first international large-scale compressor station, handed it over to the procurement and construction unit with accurate 3D models, which implemented the on-site 3D construction simulation and effectively guided the construction work [59]. Since mid-March of 2012, the "Digital Pipeline" project has been introduced into the South Xinjiang Natural Gas Favor-the-People Engineering, which was also the first "Digital Pipeline" in the Tarim Oil field. The "Digital Pipeline" project includes two aspects of pipeline digital standardization: construction and digital system platform construction. It integrates the digital platform, pipeline design, etc., and combines the global positioning system to implement unattended monitoring and patrol of the entire pipeline [60]. Based on the long-term analysis of the communication needs for long-distance pipelines in the oil and gas industry, Huawei has launched an oil and gas Digital Pipeline Information and Communications Technology (ICT) solution so as to tackle the disadvantages such as the environmental effects of long-distance oil and gas pipelines, inconvenient operations, and maintenance. It covers several aspects, namely, basic network, converged communication, integrated security, power supply, and so forth. The solution provides reliable communication Coverage, a unified and integrated office communication platform, and an intelligent security monitoring system along the long-distance oil and gas pipeline through various network Coverage modes and communication means [61].

Satellite remote-sensing technologies were employed in feasibility studies of the West-to-East Gas Pipeline, Shan-Jing 2nd Gas Pipeline, and Jing-Shen-Da Gas Pipeline. Research and practice of Digital Pipeline were carried out in the West Crude Oil Product Pipeline and the Dapeng LNG pipeline in Guangdong.

1.5 Shortcomings of Current Digital Pipeline Construction

With in-depth development of Digital Pipeline research, certain achievements of Digital Pipeline construction have been made. However, there are still many shortcomings in Digital Pipeline construction at an early stage. These shortcomings are addressed as follows:

(1) The idea of Digital Pipeline is mainly applied at stages of survey, design, and construction of pipelines. It has not been well implemented at the stage of operation management.
(2) The information acquired from the survey and design of pipelines and construction progress is not fully utilized after the design project is completed to establish an integral GIS system for information on pipelines and surroundings.
(3) Though SCADA system is widely applied in long-distance pipelines, most of them are operated on their own. SCADA system is limited in the control center and workstation, lacking effective integration with other pipeline systems. This decreases the real-time transmission and sharing of all information among

pipeline systems to a large extent, as well as narrows the scope of SCADA system application.

(4) The pipeline data model is the stepping stone of pipeline spatial database, the core of Digital Pipelines. Currently, most existing pipeline data models lack modeling for the important pipeline businesses including fire protection, repair and maintenance, pipeline real-time data, automation, fundamental geographic features, or only have limited support for these pipeline businesses.

(5) Virtual reality, an essential part of Digital Pipeline construction, has not been well implemented. The virtual reality system of Digital Pipeline can build an interactive and 3D dynamic virtual pipeline through 3D modeling of the pipeline and building pipeline virtual scenes. This provides visual application support for the operation management of the pipeline. Through the pipeline virtual reality system, pipeline users can obtain information in the way of intuitive 3D visualization. The pipeline operation can be simulated, both online and offline, under variable operation conditions. Then, the results are analyzed on the virtue occasions of the pipeline. The simulation analysis results can be displayed in the virtual pipeline scenes, which provides visual technical assistance for pipeline operation and decision-making. The Digital Pipeline virtual simulation system can also be applied to pipeline operation optimization and online operator training, etc. The pipeline network virtual reality system is an important part of Digital Pipeline, but virtual reality technology has still not been well applied in the construction of the Digital Pipeline so far.

(6) The concept of Digital Pipeline derives from that of Digital Earth, which features global interconnection and data sharing. As a part of Digital Earth, Digital Pipeline warmly welcomes a network interface for data and GIS interoperability so as to realize seamless integration with Digital Earth and other systems. However, one of the problems in Digital Pipeline construction is that the function and significance of network and distributed application are not taken into full consideration. Most of the pipelines are confined to a single machine or Local Area Network (LAN), which greatly limits its application scope.

(7) There is an insufficient combination with the latest technologies, i.e., cloud computing. New technologies, such as cloud computing, big data, Internet of Things, and artificial intelligence, have been widely applied in various industries and have produced huge economic benefits. However, the application of these new technologies in the construction of Digital Pipelines is still relatively limited. It has only been initially explored in pipeline risk analysis and internal testing and has not yet been applied in substantive fields [62]. The Digital Pipeline should enhance the combination of Digital Pipelines and these new technologies, while constantly updating its own concept and system construction.

1.6 Research Contents

In view of the under implementation of digitalization at the pipeline operation and management stage, lacking of support of pipeline data model, immature application of virtual reality technology, etc., the author of this book put forward the concept of "Web-based Digital Pipeline" and considered the trend of Digital Pipeline development toward network, current status, and demands of pipeline. Revolving around the technical architecture and development methods of Web-based Digital Pipeline, the author has put this into practice in several pipeline project such as "Yong-Hu-Ning long-distance transmission oil pipeline digitalization" and "Developing 3D GIS Web Services and Applying It in Oil and Gas Pipeline Geographical Information System".

Web-based Digital Pipeline focuses on pipeline operation and management. Its core idea lies in realization of Web-based release, query, management, sharing, and analysis of pipeline information; remote and multi-level distributed monitoring and control of pipelines; 3D visualization and virtual reality representation of pipelines, by combining with such advanced systems and technologies as Web Geographical Information System (WebGIS), GIS Web Service, pipeline SCADA, OLE for Process Control and high-speed computer network.

In the Web-based Digital Pipeline system, users can query relevant information of pipelines and surroundings areas through the WebGIS system at any terminal connected to the network, and analyze the pipeline data stored in history and real-time databases of the pipeline by the spatial analysis function provided. The Web-based Digital Pipeline realizes the virtual reality representation of the actual one. By building large-scene roaming systems and virtual facility systems of the Digital Pipeline, users are put into a visual 3D space. They explore and deal with the pipeline information through network interaction. The 3D virtual roaming of the pipeline occurs just like in the real world, which fully reflects the spatial characteristics of the pipeline. Through the construction of the Digital Pipeline visual monitoring and control system, users can conveniently carry out remote visual monitoring and control of the pipelines.

For the pipeline operation and management, the Web-based Digital Pipeline platform can provide timely and accurate data and decision-making aid for daily management and emergencies through subsystems such as WebGIS and network virtual reality system so as to guarantee operation safety. It can establish a complete database for feasibility study, survey, design, construction, production, safety, operation, and management, to realize the electronic storage of business data as well as the purpose of data accumulation for well-documented inspection. The Web-based Digital Pipeline platform provides the pipeline management departments with new means and methods through accurate and high-precision pipeline map systems and 3D visualization systems. This can contribute to the scientific, standard, and safe pipeline operation and management.

The title of volume 1 of the "Digital Oil and Gas Pipeline: Research and Practice" book series is "Pipeline Spatial Data Modeling and Pipeline WebGIS". Its main contents are as follows:

(1) The establishment of Pipeline Spatial Data Model (PSDM)

By carrying out the demand analysis of pipeline and its surrounded data, and using the object-oriented methodology, Pipeline Spatial Data Model (PSDM) is established based on ArcGIS Pipeline Data Model (APDM), and the design experience of other present main pipeline data models such as Pipeline Open Data Standard (PODS) and Integrated Spatial Analysis Techniques (ISAT).

Pipeline data model specifies the basic data structure and information coding system of the Digital Pipeline. It involves the behavior and events of the pipeline itself in construction and management, and its surrounded circumstances as well. Pipeline data model is an indispensable part of pipeline informationization construction. The rationality, validity, and openness of the pipeline data model play a key role in the construction of the Digital Pipeline. Therefore, the first task of Digital Pipeline construction is to establish a suitable pipeline data model.

PSDM is taken into full account the spatial characteristic of pipeline data, and models the properties and behaviors of relevant data along the pipeline (such as pipeline facilities, leakage, and other events), as well as the relations between pipeline spatial data and attribute data. PSDM also formulates the rules to store the spatial data in the commercial relational database, whose powerful management function can be exercised by PSDM into full play. Real-time data of pipelines is of vital significance to the pipeline operation and management, therefore, support for real-time data has to be considered in the design of PSDM. In addition, backing methods are necessary for linear referencing and dynamic segmentation since pipelines are typically linear network system. PSDM adopts module designs for pipeline elements. Several important modules such as automation, fire protection, repair and maintenance, and fundamental geographic elements are added to PSDM. Only those reusable and extensible data models are sustainable. Thus, efforts should be made to establish a versatile PSDM in its design.

(2) Pipeline WebGIS based on Web Services and Component GIS

WebGIS is a combined product of Web and GIS technology. It is a new technology developed from extending and improving GIS by Web technology. WebGIS is the best approach to implement the networked GIS application. At any node of the Internet, users can browse the spatial data in WebGIS sites, make thematic maps, and carry out various spatial information retrieval and analysis. However, a few disadvantages lie in the present main implementation methods of WebGIS. This includes the inability to achieve heterogeneous spatial data sharing and GIS interoperation, as well as cross-platform applications. To address these problems, the author of this paper proposes the research of pipeline WebGIS based on Web Services and Component GIS. The main research content is to implement pipeline GIS Web Services by using Component GIS, so as to realize the pipeline data interoperability and cross-platform applications. One of the main functions of Digital Pipeline WebGIS is to provide a pipeline information framework with the mainline of pipeline spatial location. On this platform, it can be integrated with other systems as required, and can also embed other multimedia information. Another function is to realize network-based pipeline data release, query, management, analysis, and sharing as well as interoperation.

References

1. Wang H, Guo C, He W (2005) Rising and development of digitizing pipelines. Nat Gas Oil 23(3):9–10
2. Chen J, Qin X (2005) Application of Digital Pipeline technology in survey and design field. Technol Superv Pet Ind 21(5):59–61
3. Gore A (1998) The Digital Earth: understanding our planet in the 21st century. http://portal.opengeospatial.org/files/?artifact_id=6210. Accessed 19 June 2017
4. Smith TR, Janee G, Frew J, Coleman A (2001) The Alexandria Digital Earth prototype system. In: Proceedings of the first ACM/IEEE-CS joint conference on digital libraries, June 24, 2001–June 28, 2001, Roanoke, VA, United states, 2001. Proceedings of First ACM/IEEE-CS joint conference on digital libraries. Association for Computing Machinery, pp 118–119
5. Grossner K, Goodchild M, Clarke K (2008) Defining a Digital Earth system. Trans GIS 12(1):145–160
6. Guo H, Fan X, Wang C (2009) A Digital Earth prototype system: DEPS/CAS. Int J Digit Earth 2(1):3–15
7. De La Beaujardiere J, Raskin R, Mitchell H, Rao A (2000) The NASA Digital Earth testbed. In: 8th ACM workshop on advances in geographic information systems, November 10, 2000–November 11, 2000, Washington, DC, United states, 2000. Proceedings of the ACM workshop on advances in geographic information systems. Association for Computing Machinery, pp 47–53
8. Foresman TW, Huadong G, Fukui H (2004) Progress with the Digital Earth global infrastructure. https://www.researchgate.net/profile/Hiromichi_Fukui/publication/228725007_Progress_With_The_Digital_Earth_Global_Infrastructure/links/563e0a3108ae34e98c4d7bec.pdf. Accessed 19 June 2018
9. Li D (2003) Digital Earth and "3 s" Technology. China Surv Mapp 2:28–31
10. Cheng J, Lin H, Zeng S, Zhou C (2000) Introduction to Digital Earth. Science Press
11. Guo H, Yang C (1999) Developing national earth observing system for "Digital Earth". J Remote Sens 3(02):7–10
12. Zhang H, Wang Q, Zhou C, Li H (2001) Geo-information science in the age of Digital Earth. Geo-Inf Sci 3(4):1–4
13. Chen S (1999) The "Digital Earth" as a global strategy and its master point. J Remote Sens 3(04):247–253
14. Yang C (1999) Review of the Digital Earth. Res Soft-Sci Surv Mapp 3:3–11
15. Chen S, Guo H (2000) Digital Earth and Earth observation. Acta Geogr Sin 55(1):9–14
16. Sun S, Shi P (2000) Digital Earth and its application prospects in the study of global change. Quat Sci 20(3):213–219
17. Gong H, Zhu Y, Zhao W, Zhang S, Li J (2003) Study of global change based on Digital Earth. Geogr Geo-Inf Sci 19(4):53–55
18. Tang S, Tang L, Shao G, Dai L (2006) Digital forestry research in China. Sci China Ser E Technol Sci 49:1–8
19. Shepherd M, Turner JA, Small B, Wheeler D (2018) Priorities for science to overcome hurdles thwarting the full promise of the "digital agriculture" revolution. J Sci Food Agric. https://doi.org/10.1002/jsfa.9346
20. Hu YC, Xing DL, Dong S (2014) A new digital city management based on the adaptive spatial information multi-grid technology. In: Proceedings of the Asia-Pacific conference on computer science and applications, CSAC 2014, December 27, 2014–December 28, 2014, Shanghai, China, 2015. Computer science and applications—proceedings of the Asia-Pacific conference on computer science and applications, CSAC 2014. CRC Press/Balkema, pp 435–438

21. Rathore MM, Paul A, Hong W-H, Seo H, Awan I, Saeed S (2018) Exploiting IoT and Big Data analytics: Defining Smart Digital City using real-time urban data. Sustain Cities Soc 40:600–610. https://doi.org/10.1016/j.scs.2017.12.022
22. Bremer M, Mayr A, Wichmann V, Schmidtner K, Rutzinger M (2016) A new multi-scale 3D-GIS-approach for the assessment and dissemination of solar income of digital city models. Comput Environ Urban Syst 57:144–154. https://doi.org/10.1016/j.compenvurbsys.2016.02.007
23. Jayaraman PP, Palmer D, Zaslavsky A, Salehi A, Georgakopoulos D (2014) Addressing information processing needs of digital agriculture with OpenIoT platform. In: International workshop on interoperability and open-source solutions for the internet of things, FP7 OpenIoT project held in conjunction with 22nd international conference on software, telecommunications, and computer networks, SoftCOM 2014, September 18, 2014–September 18, 2014, Split, Croatia, 2015. Lecture notes in computer science (including subseries Lecture notes in artificial intelligence and lecture notes in bioinformatics). Springer, pp 137–152. https://doi.org/10.1007/978-3-319-16546-2_11
24. Wang J-H, Wang Y-G, Fu J-H (2016) Crucial technology research and demonstration of digital mines. Meitan Xuebao/J China Coal Soc 41(6):1323–1331. https://doi.org/10.13225/j.cnki.jccs.2016.0419
25. Boxnick H (2016) The digital mine—transparency in the operational management for more efficiency and effectiveness. Die Digitate Mine - Transparenz in der Betriebsfuhrung fur mehr Effizienz und Effektivitat. World Mining Surface Undergr 68(4):247–249
26. Zhang Y, Wang C (2001) Digital valley—an important regional level of Digital Earth. Hydropower Energy Sci 19(3):1–3
27. Wu L, QingXi T (2008) Framework and development of digital China. Sci China Ser E Technol Sci 51:1–5
28. Zhao Y, Guo J (1999) The architecture research for Digital Earth. Prog Geogr 18(1):38–45
29. Milovidov KN, Gululyan AG (2016) Decision making based on digital oil field concept. In: 7th EAGE Saint Petersburg international conference and exhibition: understanding the harmony of the earth's resources through integration of geosciences, April 11, 2016–April 14, 2016, Saint Petersburg, Russia, 2016. 7th EAGE Saint Petersburg international conference and exhibition: understanding the harmony of the earth's resources through integration of geosciences. European Association of Geoscientists and Engineers, EAGE, pp 982–986
30. Chanana P, Soni TM, Bhakne U (2016) Emerging technologies and workflows in digital oil field. In: Offshore technology conference Asia 2016, OTCA 2016, March 22, 2016–March 25, 2016, Kuala Lumpur, Malaysia, 2016. Offshore technology conference Asia 2016, OTCA 2016. Offshore technology conference, pp 1487–1499
31. Ma S (2006) Building a Digital Pipeline. Digit Pet Chem 1:76–77
32. Du L, Yao A (2007) Digital techniques and its application in oil and gas pipelines. Oil Gas Storage Transp 26(6):7–10
33. Sun Q (2005) Technology of Digital Pipeline and its application. Oil Gas Storage Transp 24(4):1–2
34. Li X, Liu S, Zhang J, Fan H, Meng G, Zhang Y (2010) The status and development trend of Digital Pipeline technology. Pipeline Technol Equip 02:54–56
35. Hausamann D, Zirnig W, Schreier G (2002) High resolution remote sensing used to monitor natural gas pipelines. Earth Obs Mag 11(3):12–17
36. Brekke C, Solberg A (2005) Oil spill detection by satellite remote sensing. Remote Sens Environ 95(1):1–13
37. Roode TJ, Rector L, Smith MP (2009) Using GPS to improve data collection on the prairie waters pipeline. In: Pipelines 2009 conference, pipelines 2009: infrastructure's hidden assets, August 15, 2009–August 19, 2009, San Diego, CA, United states, 2009. Pipelines 2009: infrastructure's hidden assets—proceedings of the pipelines 2009 conference. American Society of Civil Engineers, pp 11–20
38. Longley P, Goodchild M, Maguire D, Rhind D (2005) Geographical information systems and science. Wiley

39. Li Z, Li P, Wu M, Wang W (2010) Application of ArcGIS pipeline data model and GIS in digital oil and gas pipeline. In: 2010 18th international conference on geoinformatics, geoinformatics 2010. IEEE Computer Society. https://doi.org/10.1109/geoinformatics.2010.5567619
40. Li Z, Li P, Wu M (2010) Design and realization of Digital Pipeline system based on ArcGIS engine and ArcGIS server. Comput Eng Des 31(3):638–641, 646
41. Li Z, Li P, Wu M (2009) Design and realization of Digital Pipeline system based on WebGIS. Comput Eng Appl 45(35):70–72
42. Chang G (2007) Brief discussion on Digital Pipeline construction. http://www.gongkong.com/Common/Details.aspx?c=&m=&l=&Type=paper&CompanyID=8-B9F2-1F2B4D8D438E&Id=C-A005-CE6CAF3CA963. Accessed 11 June 2018
43. Control (2008) The many faces of SCADA. Control (Chicago, Ill) 21(4):53–60
44. Hu X (2005) Virtual reality technology. Beijing University of Posts and Telecommunications Press
45. Li Z-P, Li P, Wu M (2009) Digital oil and gas pipeline visualization using X3D. In: Proceedings of Web3D 2009: the 14th international conference on Web3D technology, pp 191–196. https://doi.org/10.1145/1559764.1559794
46. Zhang X, Xue P (2006) Application of digital technique to pipeline construction. Pet Plan Eng 17(6):5–8
47. Sun X, Wen B, Tuo G (2010) Relative problems in digitization construction of long-distance natural gas pipelines. Oil Gas Storage Transp 29(08):579–581+588+554
48. Fandrich D (2014) Intelligent pipeline solution: leveraging breakthrough industrial internet technologies and Big Data analytics for safer, more efficient oil and gas pipeline operations. World J Eng Technol 02(2):55–67
49. Du Y, Yang M, Liu Y (2007) Development and application of Digital Pipeline technology. Nat Gas Oil 25(4):18–22
50. Qin G, Pang H, Liang B (2006) Preliminary study on digital technique for pipeline construction. Pet Plan Eng 17(6):1–4
51. Wang H, Liu Y, Xiao D, Li W (2007) Studies on Digital Pipeline construction in Sichuan oil field. Nat Gas Oil 25(2):1–4
52. Li S (2016) Analysis of Digital Pipeline technology standardization. Oil Gas Field Ground Eng 35(06):99–101
53. Yu T (2007) The research on Digital Pipeline design and construction. Beijing Jiaotong University
54. Han D (2006) Design management system of Digital Pipeline. Tianjin University
55. Yang M, Fu H, Du Y, Li D (2012) Digital Pipeline construction data collection. Nat Gas Oil 30(05):7–9+26+104
56. Tang J (2013) Acquisition of completion survey data for digitized pipeline during construction. Oil Gas Storage Transp 32(02):226–228
57. Dong R (2013) Constructing digital oil product pipeline system for East China based on 3S technologies. Sino-Glob Energy 18(07):83–89
58. Li H (2018) The status quo & development trend of smart pipeline technology. Nat Gas Oil 36(02):129–132
59. Yue Y, Fu Y (2018) Wisdom pipeline: unleash the unlimited value of Big Data. http://cpp.cnpc.com.cn/gdj/xwzxgdyw/201707/8278f39eaca44b43830910982a4fb856.shtml. Accessed 11 June 2018
60. Huang W (2011) South Xinjiang natural gas project to introduce "Digital Pipeline". Welded Pipe Tube 34(10):63
61. Hou S (2013) Digital journey of oil and gas pipelines. China Pet Enterp 05:59–60
62. Cheng W, Wang J, Wang X, Wang X (2018) Present conditions of China's intelligent pipelines construction and key technologies. Oil Forum 37(03):34–40

Chapter 2
Overall Architecture Design of Web-Based Digital Pipeline System

Abstract The quality of architecture design of the Digital Pipeline system directly determines the practical feasibility and performance of the system. The implementation of schemas and technical routes adopted by the Digital Pipeline system depend on the design of the system architecture. Architecture design of the Digital Pipeline system plays a coordinating, planning, and guiding role in the construction of the Digital Pipeline. This chapter introduces the architecture design of the Web-based Digital Pipeline system, including the system's objectives, design principles, functional module division, and network architecture.

Keywords Web-based Digital Pipeline · Architecture design · System objectives · Design principles · Functional modules

2.1 System Design Objectives

Based on pipeline spatial database, the Web-based Digital Pipeline system integrates advanced technologies like computer network, GIS, network virtual reality, and pipeline operation management business processes. The Web-based Digital Pipeline system aims to realize ① the release, query, management, and analysis of pipeline information based on pipeline WebGIS; ② remote and multilevel distributed monitoring of pipeline based on WebGIS through the integration with pipeline SCADA system; ③ 3D modeling of pipelines (including large-scene terrain modeling, pipeline modeling, and pipeline facility modeling) so as to realize visual monitoring and representation of pipelines. In this way, a pipeline operation management and decision-making system with high coordination, networked information exchange, and intelligent information analysis, will be formed to realize scientific, standardized, and automated management of pipeline operation. This will ultimately promote pipeline operation management levels, safe operations, and management efficiency.

Z. Li, *Pipeline Spatial Data Modeling and Pipeline WebGIS*,
https://doi.org/10.1007/978-3-030-24240-4_2

2.2 System Design Principles

To establish a practical and advanced Digital Pipeline system, certain guiding principles must be adopted. The design principles of the Web-based Digital Pipeline system are as follows:

(1) Complete: The system should be featured with complete functions, including basic functions and professional functions.
(2) Systematic: Each functional module of the system should be able to organically combine into a unified whole, and various data within each system should be easily transmittable.
(3) Practical: The system should be able to meet various needs of pipeline operation management and be able to solve user concerns.
(4) Reliable: The system reliability includes two aspects, one of which is safety and reliability of the system operation, and the other, being reliability of the data accuracy and integrity of the symbolic content.
(5) Standard: Functions of the system should meet the requirements of pipeline operation management and the information coding should follow the pipeline industry regulations.
(6) Easy to use: The system should be user-friendly; easy to use and maintain. The system should conform to business processes and information processing habits of the pipeline industry so that the requirements of operators within different levels can be met.
(7) Extensible: The system of modular structure design should be equipped with good extensibility, allowing the system to be constantly upgraded and improved.
(8) Open: The system should be able to realize the sharing of heterogeneous spatial data, GIS interoperability, and cross-platform applications.

2.3 System Functional Modules Design

The author adopts a modular structure design method for the system functions [1–3] in order to make the system extensible. The whole system is first regarded as a module, and then, gradually breaks down into several first layer modules, second layer modules, and more. One module implements a specific function, and each functional module cooperates with others to accomplish complete functions of the entire system. The function of each module should be simple and clear, with a few tasks as possible. Each module should be independent for modification and extension.

According to the characteristics of long-distance pipelines, the Web-based Digital Pipeline system is divided into three large systems, namely, pipeline spatial database system, pipeline WebGIS system, and pipeline network virtual reality system. The pipeline WebGIS system is divided into four basic modules: basic GIS manipulation module, data editing module, data analysis and processing module, and data output

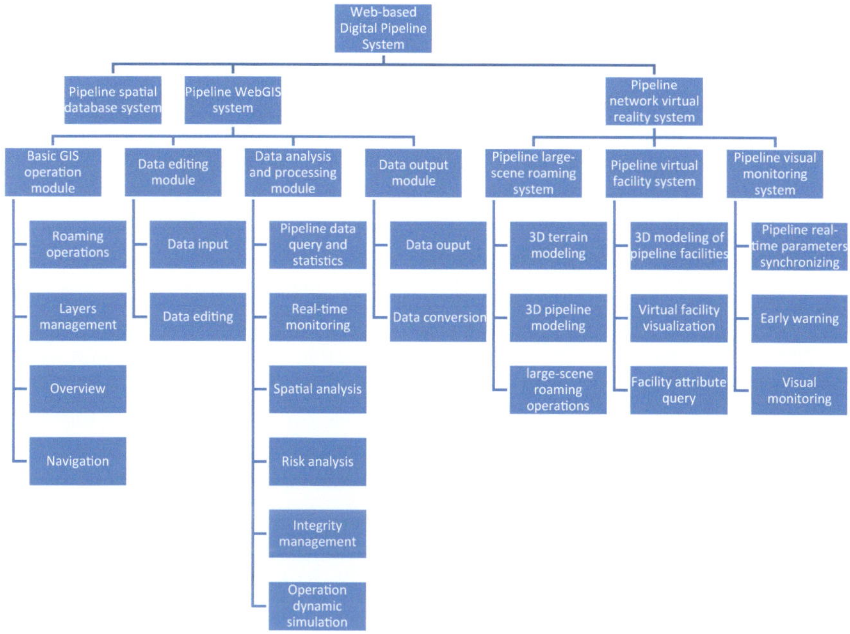

Fig. 2.1 Functional module structure of Web-based Digital Pipeline system

module [4]. The pipeline virtual reality system is divided into three subsystems: pipeline large-scene roaming subsystem, pipeline virtual facility subsystem, and pipeline visual monitoring subsystem [5]. See Fig. 2.1 for the system's functional module structure.

The pipeline spatial database system includes pipeline spatial database and auxiliary software such as ArcSDE [6]. The spatial database stores pipeline spatial data and attribute data, the digital elevation model, and high-resolution remote-sensing image map of which the main function is to provide data services for the pipeline WebGIS system and the pipeline network virtual reality system. As the function of the system is relatively clear, this book mainly introduces the design of the functional modules of the pipeline WebGIS system and the pipeline network virtual reality system.

2.3.1 Pipeline WebGIS System Functional Module Design

The major function of the pipeline WebGIS system is to realize the input, release, query, management, analysis, and output of pipeline information based on WebGIS. In addition to the basic operational functions that GIS systems should have (e.g., map manipulation such as data input, editing, and query), it is more important to have professional application functions closely related to pipeline operation management. The detailed functions of the pipeline GIS system are designed as follows:

(1) Basic GIS manipulation module: Users can perform roaming manipulations on the pipeline map to zoom in, zoom out, and pan, etc. This module also provides functions such as panorama, layers management, and overview.
(2) Data editing module: Authorized administrators can remotely edit system data by adding and deleting layers and entity features, modifying the spatial location, and attributing data of entity features.
(3) Data analysis and processing module: data analysis and processing is the core function of the pipeline GIS system. The basic function of this module is the pipe information query, including spatial information, attribute information, fuzzy query, query attribute information by spatial information, and query spatial information by attribute information. The most important function of this module is to combine pipeline spatial data and pipeline operation real-time working conditions, historical data, and pipeline professional calculation model to realize Web-based real-time monitoring, predicting, and statistical analysis of pipeline operation. This includes buffer analysis (pipeline leak prediction analysis), network analysis, route profile display, anomaly (leak, corrosion, etc.) statistical analysis, and flow analysis.
(4) Data output module: It includes functions such as map print, exporting the layer as a picture, or other format files.

2.3.2 Pipeline Network Virtual Reality System Functional Module Design

The pipeline network virtual reality system is an important part of Digital Pipeline engineering. Its principal purpose lies in establishing a Web-based, 3D dynamic virtual pipeline that can interact with users [5, 7]. On the large scale, it represents the spatial position and orientation of the pipeline on the 3D terrain model and locally reproduces the internal structure and detailed attribute information of the main facilities of the pipeline and real-time display of pipeline parameters, etc. This provides 3D visualization support for efficient management of Digital Pipelines. The detailed functions of the pipeline network virtual reality system are as follows:

(1) Pipeline large-scene roaming system: The main function of the system is to establish a multiresolution large-scale 3D terrain model based on the digital elevation model and high-resolution remote-sensing image map of the area the pipeline passes through. Then, it superimposes the 3D pipeline model on the 3D terrain model, so as to realize the large-scene roaming of the pipeline and reproduce the position and orientation of the pipeline in the form of 3D visualization.

(2) Pipeline virtual facilities subsystem: In order to enable users to understand the facilities of the pipeline more intuitively and conveniently, the system performs 3D modeling of pipeline facilities (e.g., oil stations) to realize 3D visual display of facilities and query of detailed attribute information. The details of the facilities or equipment are stored in the pipeline spatial database, and each device has a unique identifier. When the user clicks on the device, the identifier can be used to query the detailed information of the device stored in the pipeline spatial database, and to display it in the virtual scene.
(3) Pipeline visual monitoring subsystem: The purpose of this subsystem is to synchronously display the real-time parameters of the pipeline in virtual scenes, and to produce practical control effectively by manipulating the virtual scene facilities. This provides visual support for pipeline operation management. This system contains visual warning, monitoring, and other functions. For instance, as the meter readings outnumber the set range in a virtual scene, the virtual meter will flash, alarming pipeline workers to take up necessary measures. As another example, pipeline workers can press the shutoff valve in a virtual scene to stop pipeline transmission when emergencies occur.

2.4 System Network Architecture Design

In order to realize the application of the Web-based Digital Pipeline system, this book adopts distributed structure, the B/S mode for system implementation, and the three-layer network architecture (the presentation layer, the application layer, and data layer). During the process of implementing the pipeline WebGIS system, the author proposes a WebGIS implementation method based on Component GIS and Web Services in order to realize interoperability. This method contains features of the Component GIS such as powerful, flexible, and easy and convenient to develop. This method also has advantages of Web Services, i.e., good reusability, scalability, flexibility, adoption of protocol standards, and cross-platform, cross-language, and cross-hardware. See Fig. 2.2 for the architectural diagram of the entire system.

(1) Presentation layer: It mainly aims to ① provide UI services in order to be responsible for the interaction among the system and users.
It also provides a user with a friendly interface for representing information and receiving users' input, if necessary, to send users' operations to the server; ② receive and analyze the data transmitted by the server and display it for the client. Since the whole system mainly consists of two user-oriented systems (pipeline WebGIS and pipeline network virtual reality), the client configuration of these two systems is slightly different.
The client of the pipeline WebGIS system can be a standard browser or a desktop application. It is designed to provide flexibility and scalability, and to maintain a scalable interface for further development of the system. This system mainly uses a browser as a client. The browser consists of JavaScript code and ASP.NET

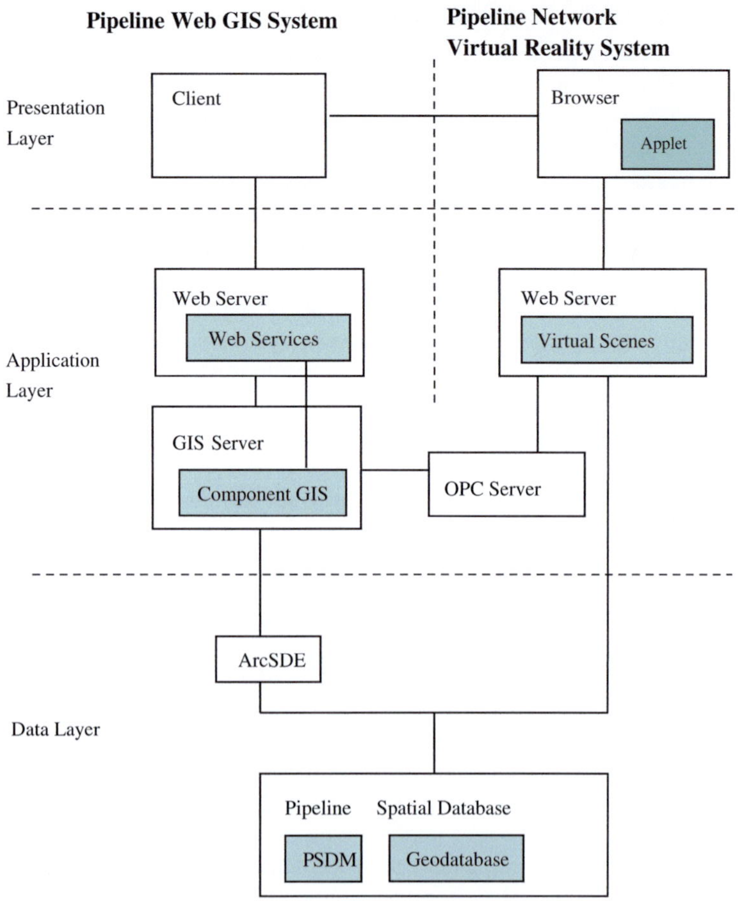

Fig. 2.2 Network architecture of Web-based Digital Pipeline system

pages, and is divided into three sub-forms: ① the graphic form displays GIS graphics, responds to various types of events triggered by the user (zoom in, zoom out, pan the map, etc.), and performs some complex functions through JavaScipt code; ② the data form displays various data queried by the user; ③ the operation form provides various tools for input, editing, and analyzing GIS data. It is unnecessary for users to install any browser plug-ins or run-times when using the pipeline WebGIS system, which is very convenient for users.

The client of the pipeline network virtual reality system is a browser, which consists of HTML pages and Java Applets. A self-developed virtual scene navigation control is embedded in Java Applets to render 3D virtual scenes downloaded from the server. The Java runtime environment is required for the client due to the use of Java Applet.

(2) Application layer: Its main role is to process various requests from customers and send the results back to them. It consists of two parts: the Web server which hosts Web Services and the GIS Server, which hosts the GIS components. In this system, the general operations of GIS data are encapsulated as Web Services interfaces, through which the pipeline WebGIS system is developed for users. It is worth mentioning that these Web Services can be reused not only by other web applications but also directly by the desktop applications. Web Services internally use the implementation method of Component GIS which are hosted in the GIS Server. The GIS Server is responsible for sending data requests to the pipeline spatial database and the GIS component carries out various spatial data analysis and processing.

(3) Data layer: Its main function is to provide data services for the system. It accepts data requests from the application layer, implements operations such as inserting, querying, modifying, updating the database, and returns the results to the application layer. The data layer consists of a pipeline spatial database and a database engine. Relevant data such as spatial and attribute data of the pipeline, DEM, and high-resolution remote-sensing image maps, are stored in the pipeline spatial database. The database engine ArcSDE, a data engine between an application and a pipeline spatial database, is used to efficiently store various spatial data stored in a relational database. ArcSDE supports multiple users, long transaction processing, and version management, and popular DBMSs such as Oracle, Microsoft SQL Server, and IBM DB2. ArcSDE solves the diversity and complexity of DBMSs and provides users great flexibility.

Within the three-layer architecture including the presentation layer, the application layer, and the data layer, there are three separate parts. These parts can be distributed on different computers in the network so as to form a distributed architecture. The layers are connected by standard communication protocols. This architecture increases the flexibility and independence of the entire system.

References

1. Wu X (2015) Geographic information system design and implementation, 3 edn. Publishing House of Electronics Industry
2. Li Z, Li P, Wu M (2010) Design and realization of Digital Pipeline system based on ArcGIS Engine and ArcGIS Server. Comput Eng Deign 31(03):638–641+646
3. Li Z, Li P, Wu M (2009) Design and realization of Digital Pipeline system based on WebGIS. Comput Eng Appl 45(35):70–72
4. Li Z, Li P, Wu M, Wang W (2010) Application of ArcGIS pipeline data model and GIS in digital oil and gas pipeline. In: 2010 18th international conference on geoinformatics, geoinformatics 2010. IEEE Computer Society. https://doi.org/10.1109/geoinformatics.2010.5567619
5. Li Z-P, LI P, Wu M (2009) Digital oil and gas pipeline visualization using X3D. In: Proceedings of the 14th international conference on 3D web technology, 2009. ACM, pp 191–196

6. Esri. What is ArcSDE? http://edndoc.esri.com/arcobjects/9.2/NET_Server_Doc/manager/geodatabase/administering_a-557706548/what_is_arcsde_qst_.htm. Accessed 06 May 2018
7. Li Z-P, Li P, Wu M (2010) Research on interaction between X3D virtual scene and Java. Comput Eng Appl 46(16):67–70

Chapter 3
Pipeline Spatial Data Model

Abstract Pipeline database is the core of a Digital Pipeline system, and the key to build a pipeline database is to design a proper pipeline data model. In this chapter, the author introduces the concepts and development history of spatial data model, and explains the design ideas, frameworks, object models, and advantages of the third generation spatial data model—the Geodatabase. The spatial data engine ArcSDE of the Geodatabase and its working modes are also described. The typical pipeline data models in the pipeline industry are comparatively analyzed from many aspects. Based on the ArcGIS Pipeline Data Model (APDM), drawing on the design experience of present main pipeline data models, combined with the actual demands and circumstances of the author's implementation of the Digital Pipeline projects, the author proposes a pipeline data model, named Pipeline Spatial Data Model, referred to as PSDM. The author elaborates on the design concepts, object models, object attributes, and modular design of PSDM, as well as, PSDM's support for linear referencing and dynamic segmentation, pipeline real-time data, and the process and methods of PSDM implementation as a Geodatabase.

Keywords Pipeline Spatial Data Model · ArcGIS Pipeline Data Model · Geodatabase · ArcSDE · Modular design · Object models · Linear referencing · Dynamic segmentation · Pipeline real-time data · Implementation

Pipeline data, featured with large volume and complex data types, is the core of the Digital Pipeline system. Throughout the life cycle of the pipeline, there are survey and design data, construction data, operation management data, and continuously generated pipeline real-time data. All of these data have spatial data characteristics, that is, pipeline data containing spatial location data. Therefore, it is difficult to manage pipeline data effectively by using traditional database methods. The disadvantages are mainly reflected in the following aspects:

1. What traditional databases manage is non-continuous, less relevant numbers and characters, while geographic data (i.e., spatial data), with strong spatial correlation, are continuous.
2. Traditional database management has fewer entity types, among which only simple fixed spatial relationships are set, while geospatial database has many

Z. Li, *Pipeline Spatial Data Modeling and Pipeline WebGIS*,
https://doi.org/10.1007/978-3-030-24240-4_3

types of entities, among which complex spatial relationships are set and new relationships are generated thereafter.

3. The data stored in the traditional database are usually atomic data recorded in equal length, while geospatial data are usually structured and have data items that can be large, complex, and variable-length records.
4. Traditional databases only manipulate and query text and numeric information, while a lot of spatial operations and queries are needed in the geospatial database, such as feature extraction, image segmentation, image algebraic operation, topology, and similarity queries.

Therefore, the primary issues of establishing the Digital Pipeline system are how to set up a pipeline data model that can effectively organize pipeline data and how to establish a pipeline spatial database that efficiently manages massive pipeline spatial data and attribute data. This chapter introduces the design and implementation of the Pipeline Spatial Data Model and the pipeline spatial database.

3.1 Research on Spatial Data Model

The pipeline spatial database is the core of the Web-based Digital Pipeline system, while the pipeline data model is the basis for the implementation and application of such system. This section will introduce the design of the Pipeline Spatial Data Model and the implementation of the pipeline spatial database.

The data model, the core and basis of the database system, is a tool to describe data content and the relationship between data and one of the main indicators to measure the capacity of the database [1]. It describes the structure of the data and defines the operations on it and relevant constraints. It also describes the static characteristics, the dynamic behaviors and constraints of the system from an abstract level, providing an abstract framework for the information representation and operation of the database system. One of the core tasks of database design is to design a well-organized data model. Therefore, the spatial data modeling has higher requirements than traditional data modeling as spatial data has spatial characteristics and the connections between data are more complex.

3.1.1 Spatial Data

Geographic data, also known as spatial data, refers to the general term for numbers, words, images, and graphics that represent the quantity, quality, distribution characteristics, connection, and regularity of the inherent features or substances in the geographical sphere or geographical environment [2]. Geographic information refers to the representation of the nature, characteristics, and state of geographic entities and all useful knowledge, as well as an interpretation of geographic data. In geographic information, the location identification is associated with data, which is the most significant sign that geographic information differs from other types of information. Geographic information is featured with regional and multidimensional structured and dynamic changes. Correspondingly, the spatial data includes three parts: spatial location, attribute characteristics, and temporal characteristics. The spatial location data describes the location of ground objects. It can be defined as geodetic latitude and longitude coordinates according to the earth reference system or, the relative positional relationship between ground objects, such as spatial distance, adjacency, overlap, inclusion. Attribute data, also known as nonspatial data, is a qualitative or quantitative indicator that describes the characteristics (nonspatial components) of certain ground objects, including semantic and statistical data. Temporal characteristics refer to the time or period when the geographic data are collected or geographic phenomena occur. Strictly speaking, spatial data is always acquired or calculated at a specific time or during a certain period of time. The temporal characteristics of spatial data may be ignored due to relatively slow changes over time. Sometimes, time can be thought of as a thematic feature.

3.1.2 Spatial Data Model

A model is a description and simulation of objects that cannot be directly observed, namely, an abstract description of complexity in the objective world. When processing real-world information with a computer, it is necessary to extract the main features of the local scope, and simulate and abstract a model that reflects the relationship between entities in the local world, namely, the data model. In other words, the data model is a tool or method to abstractly describe the real world and is a form to represent the connections between entities. It describes the data content and its connections in the database, reflecting the logical structure of the database.

The spatial data model is the abstraction and understanding of spatial data, reflection of spatial entities in the real world, and the relationship between spatial entities. It is directly related to the input, storage, processing, analysis, and output of data. It is also the foundation of spatial data organization and storage design in GIS, providing a basic method for describing the organization of spatial data and designing the spatial database pattern. This is the key to the success of design and application of GIS system.

The modeling of the spatial data is a process of transforming geographic information about real-world objects into geographic data in computers which can faithfully and objectively reflect geographic entities or phenomena in the real world, as well as their interrelationships and distribution characteristics. The data model established should naturally reflect the real world and the human being's observation and understanding of the real world as much as possible. That is, the data model should be established based on the real world and suffice the needs of users. However, from the reality perspective, the data model should be near the physical representation of the data in the computer for the purpose of easy implementation and cost reduction;

it should be oriented to the implementation and computer to a certain extent. Such requirements are obviously contradictory. In order to solve this contradiction, a multilevel data model consisting of high levels and low levels can be adopted for different users and applications. In general, the abstraction process from the real world to the computer can be divided into three hierarchical representation models: conceptual data models, logical data models, and physical data models [3].

3.1.3 Development of Spatial Data Model

The development of the spatial data model has occurred through three generations thus far. The typical representatives are the CAD data model based on file system, the Coverage data model based on combination of file and database, and the Geodatabase data model based on object-oriented technology and relational databases [4–6].

3.1.3.1 CAD Data Model Based on File System

In the early stage of GIS applications, maps were most often drawn by computer-aided design software (CAD). The CAD data model stores geographic data in binary files, and describes spatial information with vectors such as points, polylines, and polygons as well as their shapes and colors. The CAD data model also uses annotation or extended data to represent the attribute data of entities. However, it cannot store large amounts of attribute data, nor can it establish topological relationships or achieve spatial analysis. Therefore, with the development of GIS technology, the CAD data model has been gradually abandoned.

3.1.3.2 Coverage Data Model Based on the Combination of File and Database

The Coverage data model, introduced by ESRI in the United States in the 1980s, is a vector data model based on geographic interrelations. It is often considered as a geographically relevant data model and has two main features [7]:

(1) Spatial data is associated with attribute data. Spatial data is stored in binary files with index tables that optimize spatial display and access. Attribute data is stored in data tables. The number of records stored in the data table is equal to the number of spatial graphic features stored in binary files. Each record is associated with the corresponding graphic features by a common identification code.
(2) Topological relationships such as connectivity, adjacency, and region definition between vector features, are also stored. When a certain spatial feature is stored in spatial data, the node information for determining its spatial position and the

information of other lines and spatial features that are connected to or adjacent to it are stored as well.

In the Coverage data model, users can extend and define attribute tables, as well as, define the relationship between attribute tables and external databases. However, the relational database could not directly store and manage spatial data due to the limitations of computer hardware speed and database technology at the time. The idea that Coverage directly establishes and stores topological relations allowed GIS to achieve relatively high spatial data acquisition accuracy, storage efficiency, and spatial analysis ability, under the conditions of the time. Therefore, with these advantages the Coverage data model has been the standard GIS spatial data model for nearly 20 years after its roll out.

However, with the development of computer hardware and software, some of the advantages of the Coverage data model are no longer obvious, and some can be achieved by alternative and more efficient ways. For example, Coverage records spatial data in a topological structure with points, polylines, and polygons. It does not store the common sides of polygons more than once, thus saving storage space, which was an outstanding advantage in the era when the internal and external storage media were expensive. However, as the price of hardware decreases exponentially, the saving in storage space is no longer a focus of consideration. At the same time, with the development of spatial information technology and the increasing demand for GIS, the Coverage data model also shows more and more limitations, mainly in the following aspects [8, 9]:

(1) Spatial data and attribute data are stored separately in the Coverage data model. Spatial data is stored in binary files with an index table. Attribute data is stored in a standard relational database management system in the form of data tables. Spatial data is associated with attribute data by the common identification code; attribute data is based on the standard RDBMS and is supported by the efficient and reliable relational database management system. However, the data security, consistency, integrity, multiuser concurrency control, and data recovery after corruption, are not guaranteed because spatial data and attribute data are stored in different media. They have different rules, and the query operation is difficult to optimize. In addition, the Coverage data model stores spatial data in binary files, but the management functions of the file management system are weak, which limits the storage and management of massive data.

(2) Spatial data in the Coverage data model do not correspond well with its behaviors. Different types of objects in the real world are farfetchedly abstracted into simple spatial features such as "point", "polyline", and "polygon". Spatial features of the same type have the same behaviors, which makes it impossible for people to treat different entities of the same type differently. For example, in the Coverage data model, "electric pole" and "water well" can be defined as "point", so that the same operation, "moving", can be performed. In the real world, "moving pole" is reasonable, while "moving well" is far-fetched. If "pole" and "well" can be expressed in two different types of spatial features, they would have different "behaviors", and there would be no such unreasonable

behaviors as "moving well". In addition, a large number of similar examples can be cited for spatial features—"polylines" and "polygons", i.e., the Coverage data model does not support the object-oriented programming.

Although some feature models and their related behaviors can be limitedly designed through ARC macro language (AML) or VBA, the features and their characteristics are connected loosely by external codes, which make it complicated to program and are generally, error-prone. According to the "object-oriented" programming, a better approach would be to associate spatial features with their behaviors and create "spatial object" or "geographic object" models, which is exactly what the Coverage data model cannot do.

3.1.4 Geodatabase Data Model Based on Object-Oriented Technology and Relational Database

3.1.4.1 Overview of Geodatabase

With the rapid improvement of computer hardware and software, object-oriented technologies and methods have become increasingly mature, and have been widely applied to fields like computer software design, engineering, and project development. Through the object-oriented method, complex objects can be simulated and manipulated, and people's knowledge of geospatial information can be precisely reflected in computers. Therefore, the object-oriented development concept is gradually becoming the primary way of GIS development. Establishing object-oriented spatial data models is the major trend of the times.

Meanwhile, as the database technologies and functions continue to advance, it has become possible to store spatial data and attribute data directly within the same database. The Geodatabase data model is a unified, intelligent, and object-oriented spatial data model based on RDBMS launched by ESRI in this context [4, 10].

The so-called "uniformly" here refers to the model and describes the geospatial features that GIS usually processes and expresses under a common model framework in virtually the same way. The ways of processing data objects, including feature classes and datasets, and TINs and raster data, have been carried out in accordance except for the difference in collecting different types of geographic data. This helps to improve the accessibility of the Geodatabase and facilitates the customization of the Geodatabase.

The model incorporates the object-oriented core technology. The Geodatabase encapsulates all spatial features in the form of objects. It separates the external behavior semantics of the objects from the internal execution and defines the encapsulation according to behaviors. The Geodatabase structure has a strong inheritance and polymorphism. Subclasses can inherit most of the features of their superior classes, and can also have their own specific features, through which the relationship between spatial objects can be easily obtained. The object-oriented extensibility makes a predefined type of the Geodatabase data model a user type without significant changes. It can be seen that the Geodatabase data model achieves close relationships among features and behaviors at the primary level. Spatial data is no longer meaningless points, polylines, or polygons, but complex objective entity objects with attributes and behaviors for practical applications. This allows expression of geospatial features closer to people's understanding and expression of real objects.

The Geodatabase provides a scalable spatial data storage solution. It has three types of storage forms. The first type is the enterprise Geodatabase, which uses a large relational database to store data as well as a spatial data engine to manage data. It can store large amounts of data, and support long transactions, version management, and multiuser concurrent operations in a network environment. The enterprise Geodatabase is best suited for distributed GIS applications in a network environment. The second type is personal Geodatabase, which uses Access as the data storage medium. The difference between the enterprise Geodatabase and the personal one lies in that the latter has a storage capacity of up to 2 GB, can only achieve multiuser access and single-user editing, and does not support version management. The personal Geodatabase is free for ArcGIS users and is suitable for small GIS applications. The third type is the file Geodatabase, which can simulate all the information models of the Geodatabase maintained by the DBMS. It enjoys high performance, generally higher than the personal Geodatabase, and is practically equivalent to the Shapefile format. Its maximum capacity is limited to 1 TB per table. In terms of multiuser access, the file Geodatabase, like the personal Geodatabase, can support multiuser reads, but only single-user editing at a time.

3.1.4.2 Geodatabase's Description of Spatial Data

Spatial data in the Geodatabase can be separated into the following four types of datasets [4, 11]:

(1) Feature datasets: In the Geodatabase, vector data can store feature values directly in the feature dataset using dimensions and relationships. Zero-dimensional points represent relatively small geographic features, one-dimensional lines represent narrow or tortuous geographic features, and two-dimensional polygons represent broad geographic features. Descriptive vector data annotations can be used to display the names and attributes of related features. Vector data accurately describes the shapes of spatial objects through a set of sequential coordinates with associated properties and supports related spatial geometry calculations.

(2) Raster datasets: Raster data represents gridded data, including data types such as Digital Elevation Model (DEM) and image.

(3) TIN datasets: In the Geodatabase, TIN stores data as an integer of triangles with both elevation nodes and side information. Therefore, TIN can accurately describe various types of surfaces, and also support surface analysis.

(4) Datasets such as locators and addresses: Locators and addresses are special geographic representations in the Geodatabase. Locators can describe relevant address records and geographical locations in the Geodatabase, and users can find the feature values corresponding to spatial entities in the geometric network according to the locator pointers.

3.1.4.3 Geodatabase Model Architecture

The Geodatabase is an object-oriented spatial data model. In the Geodatabase, spatial entities are represented as objects with attributes, behaviors, and relations. Users can also define relationships between objects and rules for maintaining referential integrity among objects. The Geodatabase defines various types of objects such as simple objects, geographic features, geometric networks, and annotation features. These objects have complex connections and inheriting relations. See Fig. 3.1 for the core model architecture of the Geodatabase [4, 12].

The main objects of the Geodatabase are as follows:

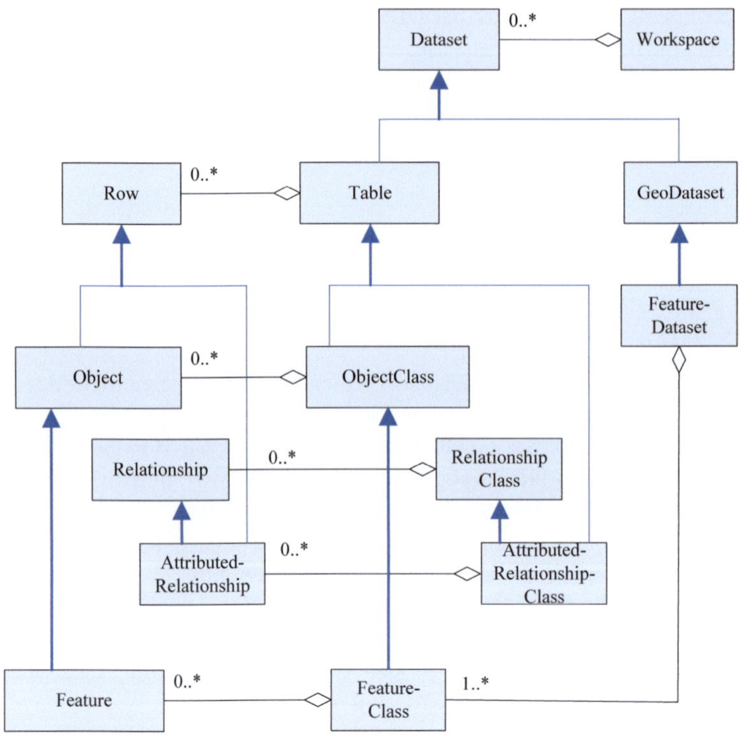

Fig. 3.1 Core Geodatabase model. Re-edited with the permission from Ref. [12]. Esri Image is used herein with permission.

(1) Workspace is a container used to store spatial data and nonspatial data. To be specific, it can be a Geodatabase, a folder of Shapefile files, or a workspace of Coverage.
(2) Dataset is an advanced container of data, which can be any collection of data, including Row, Table, FeatureClass, etc.
(3) Geodataset (geographic dataset) is a dataset containing geographic data.
(4) ObjectClass is an extension of Table. It represents a nonspatial entity with object-oriented features. It does not contain location information and is not directly displayed on the map, but instead, can be linked by a relationship class and a feature class. ObjectClass is stored as a table in Geodatabase, and an object is a row in the table.
(5) FeatureClass represents spatial entities with the same geometry and attributes. It is stored in the Geodatabase similarly to the ObjectClass, but has a special column to store spatial data. FeatureClasses can exist independently. When there are topological relationships between different feature classes, they should be organized into one dataset.
(6) FeatureDataset is a collection of feature classes with the same spatial reference.
(7) RelationshipClass is used to define the relationship between different feature classes and object classes.
(8) Object refers to an entity object with attributes and behaviors. It is a record of ObjectClass.
(9) Feature stands for an object containing spatial data and is a record of Feature-Class.

3.1.4.4 Advantages of the Geodatabase

Compared with previous data models, the Geodatabase enjoys the following advantages [4]:

(1) It supports object-oriented data modeling. Users can define their own object types due to the inheritability and extensibility of objects. They can also obtain the interaction among objects by defining their topological relevance, allowing geographic information to be demonstrated in a more natural way.
(2) It has a uniform spatial data storage mode so that all spatial data can be stored and managed intensively. Logically, the Geodatabase unifies ArcInfo's previous spatial data models and provides a uniform data interface for complex applications. All geographic data, including CAD, image, vector data, raster data, TIN, and address data, can be stored in the GeoDatabase, creating unified management, storage, and processing of various spatial data without data conversion in the same database system. Geodatabase is a spatial data model based on the relational database management system. It can integrate spatial data and attribute data into the same relational database, which removes the limitation that the two kinds of data are associated with each other only through a common identification code in the traditional models. Geodatabase can take advantage of

the efficient data management capabilities of relational databases and manage a multi-version database that supports workflow and coediting.

(3) Users can process data models more intuitively. The Geodatabase contains data objects corresponding to users' data models. The operation of the Geodatabase data not only deals with geographic geometries, such as points, polylines, and polygons, but users can also treat data like real objects they are interested in. For example, a point represents a transformer, a polyline represents a road, and a polygon represents a lake.

(4) Features are closely related. Users can define the characteristics of features and the connections between them by using topological relationships, spatial representations, and general links. When the features related to other ones are moved, changed, or deleted, the related features predefined by users will also change accordingly.

(5) The representation of spatial data is more accurate. In the Geodatabase, users can define the feature shapes via lines, arcs, elliptical arcs, and Bezier curves.

(6) It realizes the seamless storage of continuous spatial features. The Geodatabase can accommodate very large and continuous feature datasets without framing storage, block storage, and spatial separation.

3.1.4.5 ArcSDE: Geodatabase Spatial Database Engine

The digital pipe system has a huge amount of data which must be stored in an enterprise database for effective management. Geodatabase is a unified, intelligent, and object-oriented spatial data model based on DBMS. However, the object-oriented database cannot be completely fulfilled due to technical limitations. At present, the object-relational database is a relatively satisfactory solution adding a layer of management software on the top of spatial data source to realize the integrated management and object-oriented management of spatial data and attribute data. This management software is called spatial data engine and is the channel for spatial data in and out of the relational database. Its main tasks are as follows: (1) to store and to manage spatial data by using the relational database; (2) to read spatial data from the database and to convert it into the format that GIS can receive and use; (3) to import the spatial data from the GIS into the database for management by the relational database. The spatial data engine of Geodatabase is ArcSDE [13].

ArcSDE is a spatial database management software for storage of spatial data launched by ESRI. With ArcSDE, users can store multiple data products in a business relational database in accordance with Geodatabase and gain efficient management and retrieval capabilities.

The way ArcSDE stores and organizes spatial features in databases is to add spatial data types to the relational database without changing or affecting existing databases and applications. It simply includes a Shape Column in existing data tables for software management and allows access to the spatial data associated with it. ArcSDE places geographic data and spatial indexes in separate data tables and associates them with certain keys. Once a Shape Column is included in a business table, the table can be called spatial-enabled table. ArcSDE manages spatial-enabled tables by storing information in a layer table. The data in spatial-enabled tables, just like data in common ones, can be queried and merged, or queried from graphs to attributes or attributes to graphs [14].

ArcSDE stores and manages image data in the relational model database mainly by six tables: Business Table, Raster Column Table, Raster Table, Raster Band Table, Raster Blocks Table, and Raster Auxiliary Tables.

The ArcSDE vector data storage model generally conforms with the raster data storage model. It stores and manages spatial data mainly by four tables: Business Table, Layer Table, Feature Table, and Spatial Index Table.

ArcSDE enables the same function to be implemented on all DBMSs. Although all relational databases support SQL and can handle SQL in a similar way, the implementation details of different database servers are significantly different. These differences include performance and indexing, supported data types, integrated management tools and the execution of complex queries, and support for spatial data types in DBMS. Standard SQL does not support spatial data. ISO SQL/MM Spatial and OGC Simple Features for SQL extend SQL and define standard SQL support for different vector data. DB2 and Informix directly support these SQL types. Oracle uses its own standard, and its spatial type system is a separate and optional extension to the core database system. Microsoft's SQL Server does not support spatial type. ArcSDE supports the unique features provided by each DBMS flexibly. It provides additional functions that DBMSs do not have. ArcSDE addresses the diversity and complexity of DBMS, and its architecture provides users with great flexibility. Users are free to choose a DBMS to store spatial data with ArcSDE [15].

ArcSDE also provides open client development interfaces (C API and Java API), and users also have full access to the lower level spatial data tables with the applications customized by these interfaces. This flexibility contributes to an open and scalable solution, more choices, and better interoperability.

GIS data management and collection require more than a single-user large database. For any GIS system, it is more important to be able to access multiple databases, multiple formats of files, and multiple DBMSs and networks synchronously. ArcSDE can meet such critical requirements for GIS without limiting users by a DBMS or certain kind of data management solution.

With ArcSDE, the Geodatabase supports distributed applications in a network environment which greatly expands its scope of application, making it possible to build WebGIS based on Geodatabase.

3.2 Research on Linear Referencing System and Dynamic Segmentation Technology

One of the characteristics of the Pipeline Spatial Data Model (PSDM) proposed in this paper is that it supports the linear referencing system and dynamic segmentation technology. Therefore, it is necessary to introduce the linear referencing system and the dynamic segmentation technology and their applications in the pipeline industry.

3.2.1 Overview of Linear Referencing System

Geographic data can be in multiple formats according to its different representation models such as vector data representing feature collection, raster data composed of grid cells and their corresponding attributes, and triangulated irregular network (i.e., TIN) composed of a series of triangular points. Vector data is suitable for the modeling of static features such as structures and soils with fixed boundaries. However, in some applications, it is required to model relative positions of different linear features such as roads, urban streets, railways, rivers, and pipelines. For example, a leakage event occurred at a place with a distance of 61 km from the first station of the pipeline. The latitude and longitude coordinates of the leak point were 125.334735 and 43.854943, respectively. Both indicated the location of the accident: the former was located on the basis of linear referencing method, while the latter was accurately located on the basis of a geographic coordinate system. However, it is inconvenient to locate the event with the geographic coordinate system as it requires instrument assistance. Pipeline spatial information and events involved in the Digital Pipeline system are usually distributed linearly along the pipeline network. They are modeled in a 2D coordinate system (X, Y) in traditional GIS. The model built by this method can better preserve its static characteristics. Nevertheless, pipeline information is also usually dynamic, thus requiring a one-dimensional linear referencing system.

Linear Referencing System (LRS) is one of the key issues in theoretical research and application of Digital Pipelines. In the pipeline system, LRS is defined as a method for position measurements of pipeline elements with linear features in LRS space by offset from known reference points [16, 17]. It aims to integrate pipeline data with linear features and geospatial coordinates to support modeling of the linearly distributed pipeline system for practical applications such as leakage prediction analysis, pipeline flow analysis, and pipeline facility management. Therefore, LRS is of important significance for the pipeline management and operation.

3.2.2 Components of LRS

LRS consists of three basic components [18, 19]: linear referencing method, fundamental linear network, and events.

(1) **Linear referencing method**. In the linear referencing system, Linear Referencing (LR), also known as the measurement system, refers to a method of storing geographic data based on the correlation of existing linear features' positions. Position information of unknown features can be represented or measured by the position information of a known linear feature and the relative position between them [20]. Linear referencing methods include mileage reference, marker reference, and segmentation reference, but the methods suitable for a long-distance pipeline system mainly include the following:

1. Milepost referencing method: a linear referencing method for locating pipeline features by placing a milepost or a marker on the ground as a referencing point.
2. 3D projected method: control point position is obtained by selecting arbitrary x- and y-coordinate values on DEM. The z-value of the control point is equal to the elevation value of corresponding x-, y-coordinates. The distance between control points is calculated according to the following formula.

$$Distance = SquareRoot((X_1 - X_2)^2 + (Y_1 - Y_2)^2 + (Z_1 - Z_2)^2) \tag{3.1}$$

3. 3D slack chain method: control point is obtained by selecting arbitrary x-, y-coordinate values, and z-value of each control point is obtained via standard triangulation. The distance between control points is calculated by pulling a steel chain along the pipeline on the ground.
4. 3D geoid method: control point is obtained by selecting arbitrary x-, y-coordinate values. The z-value of each control point is obtained via standard triangulation. The distance between control points is calculated by the Great Circle method.
5. 2D projected method: control point is obtained by selecting arbitrary x- and y-coordinate values. It is a method based on 2D linear referencing method, so z-value is not considered. The distance between control points is calculated using the following formula.

$$Distance = SquareRoot((X_1 - X_2)^2 + (Y_1 - Y_2)^2) \tag{3.2}$$

(2) **Fundamental linear network**. The fundamental linear network is the linear spatial entities of linear referencing system consisting of routes which is a linear object with a unique identification code (route identifier) and a measurement system, such as street, highway, river, or pipeline. The route measurement is the value on the linear feature in linear reference which represents a position measured by the measurement system associated with the starting point of the linear feature.

Fig. 3.2 Point event on pipeline

(3) **Events**. Events refer to events or attributes related to the fundamental linear network, such as a valve of a pipeline, a station, or a leakage event. The leakage event is divided into point event and polyline event.

Point event describes an accurate point in the route, such as a leakage point on the pipeline and station on the bus route. A single measurement is used to describe the position of a point event. An example of a point event is shown in Fig. 3.2, in which the point event is represented by a leakage point on the pipeline.

Polyline event describes a part of the route, such as coating of pipeline, quality of road, range of salmon spawning, range of bus fares, and traffic capacity. Two measure values are used to describe positions (measure values of starting and ending points) of a polyline event. An example of a polyline event is shown in Fig. 3.3, in which, polyline event is represented by a segment of anticorrosion coating on the pipeline.

Events can be categorized and stored on a single table called event table. This includes leakage point tables and external anticorrosion coating tables. A point event table contains at least two fields: route identifier and measure value. A polyline event table contains at least three fields: route identifier, starting point measure value, and ending point measure value. Event tables can be stored in and maintained by a relational database.

The major advantage of the linear referencing method is that the pipeline information does not need to be stored in a database or added to an electronic map in geometric form (i.e., graphical), but in the form of attribute. This means that linear distribution events on the pipeline network are only represented as attributes without

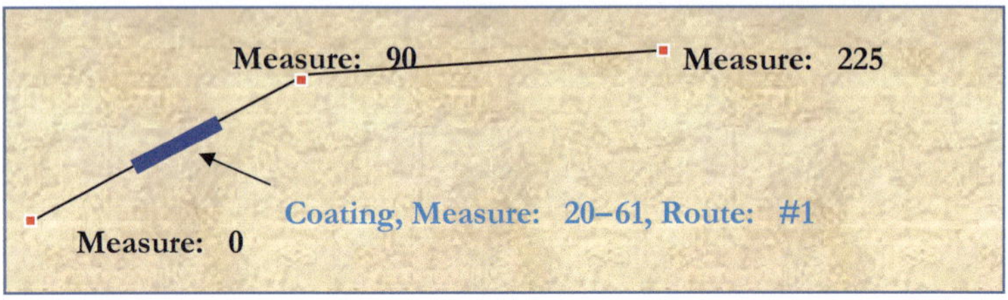

Fig. 3.3 Polyline event on pipeline

actual geographic reference coordinates. If these attributes meet the basic requirements of the linear referencing method, their spatial positions can be displayed in a geographic information system via linear referencing methods and dynamic segmentation technology [21].

3.2.3 Dynamic Segmentation Technology

In traditional GIS, linear features are stored and managed in units of arcs. While establishing a spatial database describing spatial positions of all arcs, an attribute database describing nonspatial information of these arcs is also established. At most, one record in the attribute database corresponds to each arc in the spatial database. The arc is the basic unit of attribute database with linear features, and all positions on the same arc have the same attribute values. The mode of traditional GIS processing linear features has a severe challenge in application of linear network system information. Nonspatial information about linear network system (i.e., attribute data) is collected by referring to the linear referencing system. However, linear network attribute data often has multiplicity, that is, all attributes of linear network are divided into multiple attribute sets according to certain standards to establish database, respectively, and each attribute set contains multiple aspects of information of a linear network system. The measure value of a point in each attribute database is different, which changes as attribute data changes. A major disadvantage of traditional GIS is that it can only process one attribute data set and has to process multiple attribute sets by combining them into a large attribute set in a specific way. Traditional GIS has two methods for processing the data with linear network features: the fixed segmentation method and the variable segmentation method [22, 23].

Linear features are fixedly pre-divided into several segments in advance in order to store all relevant attribute information through the fixed segmentation method. Usually, the segment should be short enough to ensure that the same attribute within each segment is consistent. The major advantage of this method is that it is simple to operate, accessible for users, and easy for software implementation. However, the disadvantage is that the data is distributed and stored with greater redundancy. Variable segmentation method divides linear features by the points where attributes change. Starting and ending points of the attribute change correspond to the starting and ending points of segments. Each segment is unequal in length because changes in various attributes are not equal in length. Data redundancy of this method is smaller making it more convenient for data maintenance and updates. Nevertheless, this method requires that linear features are segmented according to each type of attribute. Spatial positions of linear features have to be divided repeatedly by different attributes, resulting in repeated storage of spatial data. The spatial position of the same linear feature has to be repeatedly inputted. In view of limitations of above methods, many scholars have studied to establish connections between various attribute sets and spatial positions of linear features without changing the linear feature's spatial data. This does not spatially segment linear features, but dynamically displays certain attributes of linear features in segment according to the established linear referencing system when querying and analyzing. Thus comes the idea of dynamic segmentation.

Dynamic segmentation uses the linear referencing system and corresponding algorithms based on the traditional GIS data model to dynamically calculate the spatial position of the attribute data for analysis, display, query, and output [24]. Dynamic segmentation solves problems encountered by traditional GIS in processing linear features. It is a new technology for dynamic analysis, display, and mapping of linear features, and can greatly improve the processing of linear features [25]. The emergence of dynamic segmentation has opened up vast prospects for the application of GIS within the road, railway, river, and pipeline fields.

Dynamic segmentation has the following characteristics [22]:

(1) It realizes dynamic display and analysis of multiple attribute sets without repeated digitization leading to less data redundancy.
(2) Linear features are not actually segmented according to attribute data sets. Various attribute data sets can be dynamically displayed in a segment for analysis and query when necessary.
(3) All attribute data sets are based on the position description of the same linear feature, that is, attribute data organization is independent of the base map of the linear feature for data updates and maintenance.
(4) It allows comprehensive query and analysis of multiple attribute data sets.

Dynamic segmentation involves three aspects of data processing of GIS [26]:

(1) Connection: This refers to establishing connections between attribute information corresponding to distance or position and entities that are linearly distributed in geographical space.
(2) Segmentation: This refers to the process of generating new point entities or polyline entities based on selected attribute data and certain conditions in analysis and query.
(3) Display and spatial analysis: This refers to accurately representing generated point entities or polyline entities in 2D space according to the results of connection and segmentation for further spatial analysis.

3.2.4 Dynamic Segmentation Algorithm

The core of the dynamic segmentation is the segmentation algorithm. There are two main algorithm models currently used. One is the interpolation method [27] by which the coordinate of an uncertain point is generated based on the coordinates of measure values of known points, and measure values of unsolved points. The algorithm flow of this method is relatively simple. Another dynamic segmentation algorithm is based on route curve features [23] and is calculated according to alignment and parameters of the route making the algorithm flow more complicated. The interpolation method is adopted for dynamic segmentation technology in this book and considers moderate curvature change of the pipeline. Flow of the algorithm is illustrated in the following example.

In Fig. 3.4, each control point and reference point are known, and measure values of the two points, B and D, are known as well. The task is to generate B and D through the dynamic segmentation method as well as, highlight the segment BD. The algorithm flow for solving this example is shown in Fig. 3.5.

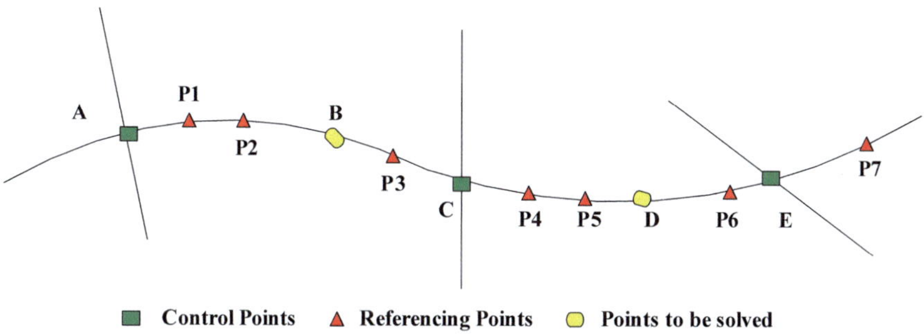

Fig. 3.4 A diagram illustrating dynamic segmentation using the interpolation method

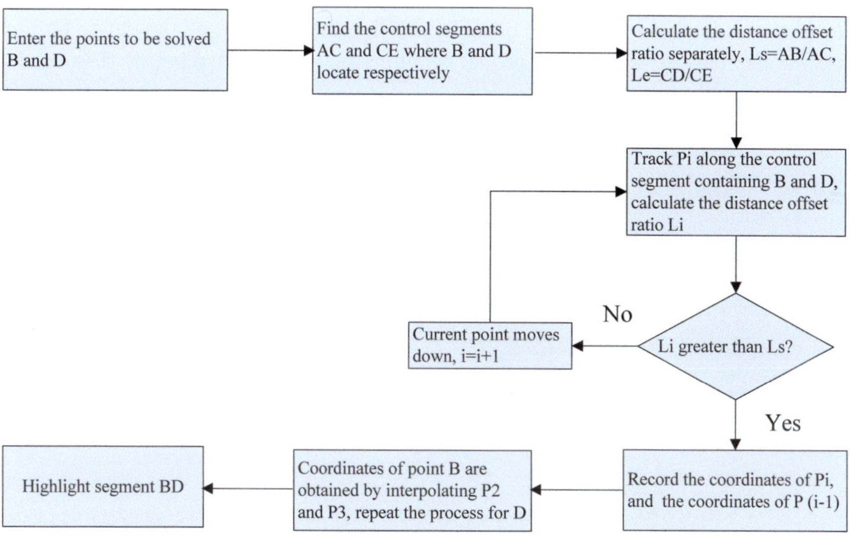

Fig. 3.5 Algorithm flow of dynamic segmentation using the interpolation method

3.2.5 Application Examples of Linear Referencing and Dynamic Segmentation in Pipeline Analysis

In addition to the dynamic display of pipeline information, query and analysis of pipeline operational data and historical data are more important functions for introducing linear referencing and dynamic segmentation in the pipeline spatial data model. For example, a task is to carry out a statistical analysis on leakage accident occurring on seriously corroded segments, of which the steps are as follows:

(1) Display leakage points on the pipeline by using dynamic segmentation technology. The result is shown in Fig. 3.6.
(2) Display seriously corroded segments on the pipeline by using dynamic segmentation technology. The result is shown in Fig. 3.7.
(3) Number of leakage points in severely corroded segments is obtained by combining the above two results. The result is shown in Fig. 3.8.

3.3 Comparative Analysis of Pipeline Data Models

The pipeline data model is a data model applied in the oil and gas pipeline field. The pipeline database is the core of a Digital Pipeline system, and the key to build a pipeline database is to design a proper pipeline data model. A pipeline data model

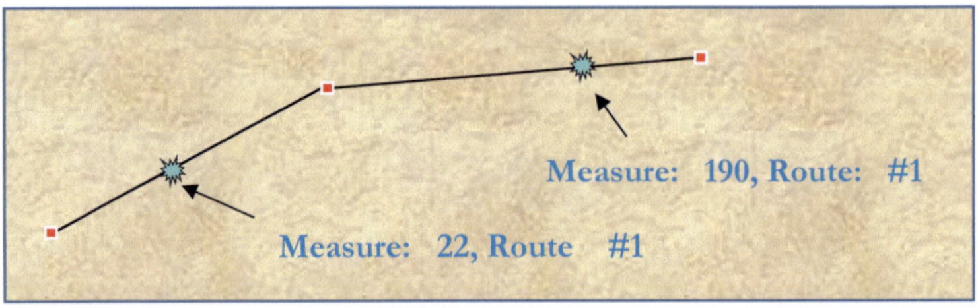

Fig. 3.6 Query result of pipeline leakage points

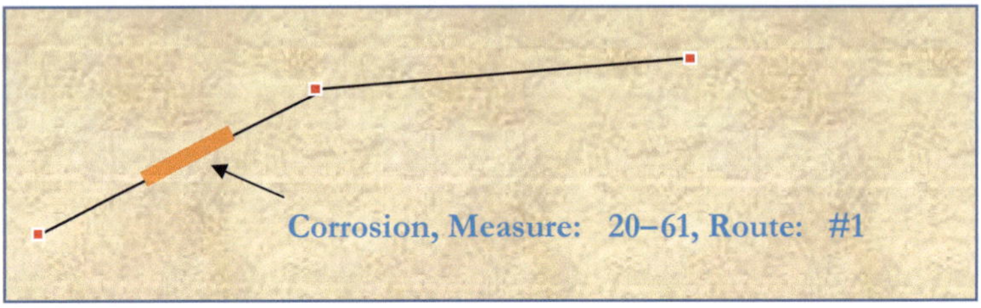

Fig. 3.7 Query result of seriously corroded segments

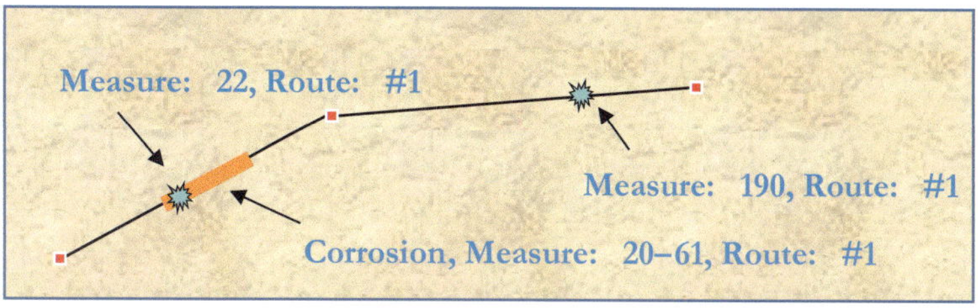

Fig. 3.8 Query result of leakage accidents occurring on seriously corroded segments

provides a framework for gathering, organizing, and managing pipeline information. The roles and benefits of pipeline data models are listed below [28]:

- Provide a uniform data structure, database, and comprehensive data inventory for data required for the Digital Pipeline.
- Improve data accuracy, consistency, and integrity, all of which are essential for Digital Pipeline information management.
- Efficiently manage pipeline data in an ordered, centralized, and interrelated way.
- Easy, fast, and effective to update pipeline data.
- Continually provide high quality and reliable data.
- Provide interoperability between pipeline companies and better support of integration activities, as well as promote data sharing with industrial standard data models.
- Be able to implement and customize according to own need with the flexibility and extensibility of the pipeline data model.
- Provide spatial support for pipeline GIS implementation.
- Reduce cost, time, and complexity of GIS implementation.

Many pipeline companies have started research on pipeline data models in the 1990s. So far, representative pipeline data models around the world are as follows: Integrated Spatial Analysis Techniques (ISAT), Pipeline Open Data Standard (PODS), and ArcGIS Pipeline Data Model (APDM). In recent years, many departments and scholars in China have also studied pipeline data models. Jin Jian et al. established Pipeline Automatic Data Model (PADM) [29] by referring to PODS and APDM. Cheng Zhongyuan proposed Chinese Pipeline Data Model based on the APDM [30]. Jia Qinglei et al. proposed APDM-CH based on the APDM [31]. PetroChina pipeline science and technology research center developed the Pipeline Integrity Data Model (PIDM), [32] etc. However, pipeline data models proposed in China are mainly based on refinement or extension of the above three models due to the studies having started late. The following is a brief summary of the three representative pipeline data models mentioned above.

3.3.1 ISAT

In 1994, the American Gas Technology Institute (GTI), a well-known energy research institute, recognized the importance of establishing a data standard covering the entire pipeline industry. They funded the development of Integrated Spatial Analysis Techniques (ISAT), which was a milestone for the study of pipeline data model. The pipeline industry also has the first data model which does not require the pipeline operators to customize the pipeline data every time. ISAT was designed for [33]:

(1) Managing regulatory inquiries,
(2) Improving decision-making process,
(3) Improving quality and efficiency,
(4) Reducing human resources, and
(5) Developing new technologies.

Pipeline operators can obtain the following benefits by using the ISAT pipeline data model [33]:

(1) Avoiding custom database design,
(2) Reducing application development costs, and
(3) Reducing data migration costs.

After 3 years of hard work, the ISAT project team released the first pipeline facility data model, named ISAT, as well as the first program developed by M.J. Harden Associates to manage pipeline data called the PipeView.

ISAT mainly models the following objects [33]:

(1) Pipeline hierarchy, routing, and physical position of pipeline facilities;
(2) Pipeline stations;
(3) Pipeline equipment (pumps, compressors, valves, conduits, etc.) information; and
(4) Pipeline encroachments.

ISAT can record events and information related to the above objects, including:

- Preventive maintenance,
- Condition monitoring (temperature, pressure, flow, cathode test point readings, rectifier readings, etc.),
- In-line inspection,
- External inspection,
- Non-destructive testing (X-ray, X-ray photography),
- Pipeline cleaning,
- Engineering analysis,
- Maintenance,
- Risk assessment and management, and
- Geotechnical investigation.

ISAT was designed for promoting pipeline management and maintenance efficiency, while paying less attention to the pipeline design and engineering. It includes a set of core tables modeling basic pipeline operational data and facility data required for routine management of pipelines and can be extended by pipeline operators for specific needs. In addition, ISAT is based on the relational data model. Its core sets are independent of specific databases and programs, which allows ISAT to be implemented as a database system compatible with any SQL.

One major disadvantage of ISAT is that it does not provide built-in support for GIS functions. Therefore, a column that provides a link between vector (or image) features and pipeline attribute data should be added to the ISAT table to ensure that the pipeline information system using ISAT can support GIS. That makes the model increasingly unable to meet the growing demand of the pipeline information system for GIS.

3.3.2 PODS

GTI planned to rewrite or extend ISAT to overcome its shortcomings. Its initial goal was to establish an open and exchangeable pipeline data standard, so the model was called Pipeline Open Data Standard (PODS). GTI set up a working group for this work who added a lot of functions to the original ISAT and greatly expanded its scope.

Compared with the ISAT pipeline data model, PODS has been extended in the following areas [33]:

(1) Adding support for liquid pipeline information;
(2) Adding more modeling on pipeline facilities and more detailed information;
(3) Supporting multiple coordinate references;
(4) Containing network tables to store topology information;
(5) Supporting event groups for organizing events in a hierarchical manner;
(6) Supporting pipeline operation and status monitoring information;
(7) Supporting inspection events, including in-line inspection, external inspection, remote inspection, excavation, and direct inspection;
(8) Regulatory compliance events;
(9) Risk assessment event;
(10) Historical data events; and
(11) Supporting linear reference.

PODS was designed for [33]:

(1) Improving the integrity of pipeline data,
(2) Reducing the risk of implementation,
(3) Establishing an industry data standard,
(4) Improving data interoperability, and
(5) Improving data management efficiency.

PODS organization was established in August 2000. By 2018, there have been more than 160 members in the PODS Organization, including pipeline operators, pipeline service providers, government agencies, and associations, which are organized together to help define and implement the pipeline data standard. The version 6.1 of PODS was released in 2018, and has already included about 1,000 tables covering pipeline positions, pipeline facilities, geographic features, in-line inspection, cathodic protection, physical inspection, risk, offline events, and other contents [34].

PODS has the following characteristics [35]:

(1) It covers various aspects of pipelines.
(2) It is mainly used for managing pipeline business and integrity but contains comprehensive pipeline-related information.
(3) It is an open standard.
(4) PODS Association owns its intellectual property.
(5) It is designed by members of PODS.

PODS is also defined as a model based on the relational data model which is similar to ISAT. The core set of PODS is independent of a specific database and program. This enables it to be implemented as a database system compatible with any SQL, including Oracle, SQL Server, Sybase, and Informix.

Event table is the core of PODS and is linked to the event attribute table and the feature table. Event attribute refers to an event or entity related to pipelines, such as a leakage event and a valve. Feature table stores the type and position of event features. Coordinate values can be stored by spatial geometric features, but is independent of specific GIS in order to maintain the openness of PODS. Event, event attribute, and feature are linked by an identical identifier. Event table also stores historical data for data backtracking which is a very important function added by PODS but may also lead to accumulation of a large amount of data. It is recommended by PODS to store the current data and the historical data separately, i.e., within different databases.

The database and software based on PODS must comply with PODS compliance standards [36]. PODS is extensible, but it is a controlled process. PODS Association has developed a set of standards that pipeline operators and service providers must comply with in order for implementation of the PODS-based database and software, and extension of PODS. The set of standards are based primarily on data interoperability, upgrades, document integrity, compatibility, and stability. The set of standards are summarized as follows:

(1) Any database object (including tables, columns, and relationships) of PODS cannot be deleted.
(2) New tables and columns can be added to PODS as long as they conform to the design ideas and concepts of PODS.
(3) Database objects, such as triggers and stored procedures, can be added to PODS as needed.

PODS is designed for developing and promoting the standard for data exchange and storage in the pipeline industry. Its goal is to provide a public platform for pipeline companies to build a standard-based GIS database. PODS seeks to represent the needs of the entire pipeline industry by providing a public framework that allows pipeline companies to focus on their specific needs rather than general needs of the pipeline industry. Pipeline companies are free to choose technologies provided by multiple service providers due to interoperability and compatibility of PODS, and can also replace service providers as needed without rebuilding the database [35, 37].

3.3.3 APDM

The APDM is a data model for storing information related to oil and gas pipelines. It is explicitly designed in the way that the APDM is implemented using the Geodatabase and its GIS functions are implemented using ArcGIS series products [38]. It is different from PODS which is independent from specific database and GIS.

In March 2002, M.J. Harden Associates Inc. launched the research on APDM. Since then, many pipeline operators have participated in the discussion and design of the model. At ESRI Petroleum User Group Conference held in March 2003, APDM was released for public comments. At ESRI Petroleum User Group Conference held in July 2003, APDM version 1.0 was officially released. The latest version of APDM, version 6.0, was released in 2013.

APDM is derived from existing released pipeline data models (including PODS and ISAT) and has extended to meet the needs of the modeling oil and gas pipeline data. It is designed and developed by ESRI Pipeline Interest Group Steering Committee and Technical Committee under the guidance of ESRI. Members of the Technical Committee include representatives of pipeline operators and pipeline service suppliers. APDM is designed to include 80% of the most common and standard contents in pipeline industry, particularly in pipeline industry's most concerned areas such as integrity management, pipeline inspection, high consequence areas, and risk analysis. It is not an extensive data model or a model covering everything, but a template on which pipeline operators can start with core features of APDM and modify the data model by adding or refining features [38]. Another major design objective of APDM is to support linear referencing system. Most pipeline companies use measure values to locate pipeline features or events occurring along the route [39, 40].

APDM is designed as the starting point for pipeline data modeling. It aims to provide a set of core objects and a set of attributes to describe and effectively support linear referencing. It also aims to provide a set of core conceptual objects to summarize most of the features of the pipeline system, rather than to describe all of the features of the pipeline system or specify a rigorous or standard method for data modeling of the piping system. The purpose of the set of core objects provided by APDM is to provide pipeline service providers with a consistent framework to develop APDM-based applications and transform data between existing databases. Any pipeline company can add features (except the set of core objects) to APDM, modify existing features in APDM, or even remove unnecessary features [41].

Another goal of APDM is to provide pipeline companies with ESRI's core technologies for developing APDM-based piping applications.

Design goals of APDM are as follows [40]:

(1) Reducing the costs of companies by establishing a data model that can be extended as needed,
(2) Reducing expenses required for software development,
(3) Supporting positioning pipeline features by two methods of linear referencing and geographic coordinates,
(4) Using mature technologies of ESRI, and
(5) Acting as the starting point of the pipeline information system.

APDM intends to be a flexible data model template which can be modified by any pipeline company to meet its needs. If the pipeline company has established a traditional pipeline relational database, it can also transfer that database to APDM to make full use of ESRI's Geodatabase data management environment. If the pipeline company has not yet established a pipeline database, it can also use available classes of APDM to store pipeline information.

3.3.4 Comparison of PODS and APDM

According to the investigation of 56 North American pipeline companies in 2005, the proportion of common pipeline data models usually used is shown in Fig. 3.9 [42]. From the figure, ISAT still occupies a large proportion. However, PODS contains the ISAT contents and is supported by an increasing number of pipeline companies. The proportion of ISAT will gradually decline. APDM provides powerful built-in GIS concepts and spatial database, and is customizable making it likely to be more widely used. Therefore, PODS and APDM will become the two most commonly used pipeline data models. PODS and APDM have their own advantages. Their similarities and differences are summarized as follows.

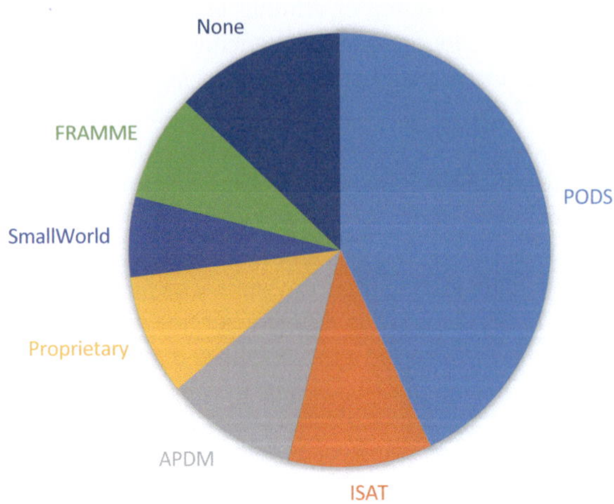

Fig. 3.9 Proportion of pipeline data models used based on the investigation of 56 pipeline companies in North America

Similarities between PODS and APDM [43, 44]:

- Supporting the data modeling of similar pipelines—oil and gas pipelines;
- Supporting positioning pipeline features by two methods of linear referencing and absolute positioning;
- Managed by a Technical and Management Committee composed of pipeline operators and service providers;
- Supporting GIS functions;
- Choosing solutions from service providers;
- Integrating solutions from multiple service providers; and
- Providing technical support through training, seminars, user group meetings, etc.

Differences between PODS and APDM [43, 44]:

- APDM is a template with a standard core.
- The standard core of APDM must be implemented as required to achieve maximum interoperability of databases and commercial software.
- The remaining classes of APDM, except the core classes, are optional and are the best practice results in the pipeline industry. If chosen, those classes must be implemented by following the rules of APDM abstract classes.
- PODS is a standard designed to define pipeline features with rich details to describe pipeline information.
- The entire PODS table structure must be implemented as originally defined, which is an important part for a "standard" PODS.
- APDM can only be implemented as ESRI's Geodatabase and also has built-in support for GIS functions. Its implementation process is standard, usually involving the same technologies (linear referencing, topology, etc.).
- PODS is a relational database management system without built-in support for GIS functions. It stores linear referencing and coordinates in tables.

A conclusion can be made by comparing the two models:

PODS is an independent model for GIS, which provides pipeline operators with the freedom to choose the most appropriate GIS application. APDM uses the technology provided by ArcGIS and also provides pipeline operators with the flexibility for implementation.

Although PODS and APDM are two pipeline data models with different positioning and definitions, there is no competition between the two. On the contrary, they cooperate with each other in an effort to provide the best implementation solution for pipeline operators and service providers. PODS Association and APDM Association have reached a formal agreement, the main content of which is to implement PODS ESRI Geodatabase [45].

3.4 Design and Implementation of Pipeline Spatial Data Model

Based on APDM, drawing on the design experience of present main pipeline data models of the pipeline industry, combined with the actual demands and circumstances of author's implementation of the Digital Pipeline projects, the author proposed a pipeline data model, named Pipeline Spatial Data Model (PSDM). The design and implementation of PSDM are described in detail in the following sections.

3.4.1 Features and Advantages of PSDM

The design of PSDM is based on the schema of APDM. The design concepts of PODS and ISAT are also taken as important references. PSDM emphasizes spatial characteristics of pipeline data and fully considers the actual situation of domestic pipelines. The core object model architecture of PSDM is basically consistent with that of APDM [40], but the following major improvements and extensions have been made:

(1) PSDM supports real-time parameters of the pipeline operation, which refers to pipeline real-time operating condition data. This includes pressure, temperature, flow, water cut, and density of a certain point on the pipeline, and the inlet and outlet of a station; pressure and temperature of the oil pump inlet and outlet, valve states; and electrical parameters of the oil pump motor. Pipeline real-time parameters are vital for realizing pipeline automation, as well as, the data source of various pipeline analysis applications software including pipeline dynamic prediction simulation, pipeline operation optimization, leak detection, and positioning, real-time simulation, and surge dynamic analysis. Therefore, pipeline real-time parameters are of great significance for implementing integrity management, ensuring pipeline economical and safe operation, and providing decision support. However, the current existing pipeline data model generally lacks systematic support for pipeline real-time parameters. PSDM supports pipeline real-time parameters, and solves problems in pipeline real-time parameter acquisition, display, integration, etc. This is a major breakthrough in pipeline data modeling and has important practical engineering significance.
(2) PSDM has a modular design of pipeline elements. According to different business roles and functions, pipeline elements are divided into such modules as pipeline Centerline facilities, sites and facilities, measurement and testing, crossing defects and inspection, cathodic protection, fundamental geographic features, risk and protection, fire protection, repair and maintenance, and automation. With modular design, the use of the model becomes more flexible and the extensibility of it is also enhanced. After the implementation of the model to a GIS, it is convenient to switch the display of the overall element of a module. New modules can be added and unwanted modules can be removed according

to the actual demands of the project. A module can also add pipeline elements or remove unwanted elements.

(3) The pipeline automation system (SCADA system) is added to the model. The pipeline automation system monitors and controls the running equipment on-site to realize various functions such as data collection, equipment control, measurement, parameter adjustment, and various signal alarms. It can complete real-time collection and transmission of on-site data, as well as provide local or remote control on the industrial site, perform comprehensive and real-time monitoring of the process flow to provide the necessary data and basis for production, and schedule and manage. The pipeline automation system plays an important role in ensuring safe operation and optimization of pipelines, increasing efficiency, and improving the working environment. However, the current pipeline data model lacks effective support for pipeline automation system modeling. PSDM abstracts and generalizes the production automation business involved in pipeline operation. It conducts spatial modeling on the main objects in the field of pipeline production automation, including station control host, PLC, RTU, and communication network. This gives geographic reference coordinates to these objects, which makes the integration of the pipeline automation system and the pipeline GIS possible, and at the same time, lays a good foundation for the integration of management and control.

(4) The modeling of fire protection and pipeline repair and maintenance module are added to the model. Oil and gas transmitted by pipelines are generally flammable and explosive, and some oil and gas media contain toxic and harmful components. In the event of accidents such as leakage, not only can it cause serious waste of resources and economic losses, but also, environmental pollution and even fires and explosions resulting in threats to people's lives and property, and the living environment. With the increasing requirements of safe production of pipelines from government, enterprises, and the public, it is necessary to strengthen and improve the management level of pipeline fire protection and repair and maintenance. PSDM models relate elements of pipeline fire protection as well as repair and maintenance businesses (including facilities, equipment, and locations). Through the integration with other pipeline business systems such as pipeline GIS, it will be able to realize the visual management and scheduling of fire protection and repair and maintenance, thereby improving the ability to quickly respond and deal with emergencies and ensure safe construction and production of the pipeline.

(5) There is added modeling of fundamental geographic features. Fundamental geographic features include high-resolution remote sensing images, digital elevation models, geological conditions, roads, railways, rivers, administrative divisions, residential areas, and factory areas within a certain range along the pipeline. Fundamental geographic features play a very important role in the pipeline survey and design, route selection, rerouting, high consequence area analysis, and evaluation, emergency disposing planning, and pipeline operation analysis.

(6) PSDM modifies or deletes classes in APDM that does not meet the actual situation of the domestic pipeline.

3.4.2 PSDM Design Principles

PSDM design principles are as follows:

(1) Aims at implementing as an ESRI Geodatabase: As stated above, PSDM is a spatial data model on the basis of object-oriented technology and relational database, with advantages of supporting object-oriented, and providing uniform spatial data storage, intelligence, and seamless storage of continuous spatial features. Therefore, in order to utilize these advantages of Geodatabase, the design concept, data structure, and logical structure of Geodatabase in the design process of PSDM are fully made use of.
(2) Generality and extensibility: In order to realize data sharing in the pipeline industry and improve interoperability of pipeline data, PSDM is designed to be a general and extensible pipeline data model. Based on the demand analysis of pipeline data, PSDM contains most of the pipeline industry elements, while the data not defined by PSDM can be extended by the pipeline company with the extensibility of PSDM. The extensibility of PSDM is achieved through the object-oriented inheritance mechanism.
(3) Supports linear referencing and dynamic segmentation: The positioning of features on the pipeline can be solved with the absolute positioning method and relative positioning method. The absolute positioning method refers to locating objects by their geographic coordinates, while the relative positioning method refers to locating objects by linear referencing system. Although the absolute positioning method adopted for pipeline features is increasingly accurate with the development of GPS, GIS and RS, the relative positioning method for pipeline features through linear referencing still has its unique advantages. PSDM provides support for both positioning methods. At the same time, it also supports dynamic segmentation of pipeline features on the basis of linear referencing.
(4) Supports real-time data: Real-time data of the pipeline system, including pipeline temperature, flow, pump station inlet, and outlet pressure, are of vital importance to the operation management of pipelines. A major characteristic of PSDM is to support real-time data of pipelines. Pipeline SCADA systems, generally installed in today's pipelines, are the source of PSDM real-time data. Therefore, features in PSDM containing real-time data are mainly facilities and equipment used for real-time monitoring in the pipeline SCADA system.
(5) Suitable for domestic practical situation(s): There are also many differences in requirements of pipeline data modeling due to different conditions of domestic and foreign pipeline industries. Pipeline data models such as ISAT, PODS, and APDM mostly refer to pipelines in North America. Therefore, there are many aspects that are inconsistent with domestic conditions in these pipeline data models, such as compliance. In the design process of PSDM, the specific condition of the domestic pipeline industry has been fully considered and strives to make PSDM a pipeline data model suitable for domestic pipeline data modeling.

(6) Comprehensiveness: It should cover, as much as possible, the useful data generated throughout the life cycle of pipelines, including pipeline survey and design data, pipeline construction data, online inspection data, and real-time monitoring data after pipeline commissioning, so as to provide a source of information for pipeline operation management decisions and lay a good foundation for the establishment of pipeline integrity management and the risk assessment system [46].
(7) Accuracy: The relations between pipeline features should be reflected as accurately as possible. Those relations include attachment and containment of pipelines and their facilities, as well as, online and offline of pipeline facilities to pipelines.

3.4.3 *Linear Network of PSDM*

In the design concept of PSDM, the collection of all of the main lines and branch lines of a pipeline system is collectively referred to as Centerline. Centerline consists of a series of StationSeries, which acts as a route of linear referencing system. Each StationSeries corresponds to a linear referencing system, and can use different linear referencing methods. A StationSeries object will be implemented as a polyline feature in spatial database, describing the physical direction of the pipeline.

A StationSeries object is created by connecting ControlPoints. A ControlPoint object represents a point of known geographic reference coordinate and measure value along a StationSeries, and each ControlPoint corresponds to a three-pile (mark pile, corner pile, or milepost). At each turning point of StationSeries there is a ControlPoint, forming the vertex of StationSeries and controlling the direction of StationSeries. In this way, at least two ControlPoints are required to form a StationSeries. ControlPoint is further divided into two types: node, the start or ending point of a StationSeries, and internal point, the point between the start and ending point of a StationSeries. Each ControlPoint can be involved in the construction of different StationSeries allowing one ControlPoint to have different measure values.

StationSeries and ControlPoint must be implemented in PSDM as feature classes, i.e., features with geometric fields. Meanwhile, a StationSeries and all ControlPoints that make up the StationSeries must share the same linear referencing system. Measure values at other points between ControlPoints are obtained by linear interpolation.

StationSeries forms the basic linear network of the pipeline linear referencing system. Linear events or facilities occur or exist on StationSeries, that is, events or facilities depend on StationSeries.

3.4.4 PSDM Feature Classification and Modular Design

Objects in PSDM can be divided into online features and offline features. The difference between online features and offline features is that the online feature object is located on Centerline and positioned by linear referencing system, while the offline feature object is located by geographic reference coordinates.

Each online feature in Centerline has a measure value, which takes a point of Centerline as the benchmark. For the online point feature, only one measure value is required to locate the feature, while the online polyline feature requires a starting measure value and an end measure value to fully locate the feature. Although offline feature is positioned with geographic reference coordinates, connection between offline features and Centerline can also be created through predefined relation classes.

Classes in PSDM can be divided into Object Class and Feature Class according to whether they have geometric attributes, that is, whether they contain spatial data. Object class does not contain geometric attributes while Feature Class does.

In order to realize interoperability and extensibility of PSDM, PSDM is designed as a three-level object model structure of Abstract Classes, Core Classes, and Entity Classes.

Abstract Classes define the framework of PSDM. An Abstract class defines the common characteristics of a pipeline feature class. Each nonabstract class inherits attributes, relations, and behaviors from an Abstract class. Core Classes define Centerline features, linear reference, and other key features of PSDM. Entity Classes, inheriting from Abstract Classes, are used to describe specific features, such as oil transportation station, pump, and valve.

As mentioned before, for the reason of promoting flexibility and extensibility of PSDM, PSDM is divided into 12 modules including the Essential module, the Automation module, and the Measurement and Testing module. Among these modules, the Essential module, consisting of Abstract classes and Core classes, is the only required module in PSDM, while the other 11 modules are optional modules. Each optional module is also referred to as Entity modules as it is composed of Entity classes. The POSDM modules diagram is shown in Fig. 3.10.

3.4.5 PSDM Hierarchy Design

Pipeline companies often need to organize the pipeline and its features hierarchically for convenient management reason. The hierarchy is usually divided based on where StationSeries is located. Here is a typical organized hierarchy: multiple StationSeries are organized into one LineLoop, the multiple LineLoops are organized into one pipeline system, at last multiple pipeline systems are managed by a pipeline company. Even if a simple pipeline system can be divided into more complex hierarchical structures, such as main lines, branch lines, and sub-transmission subsystems, for

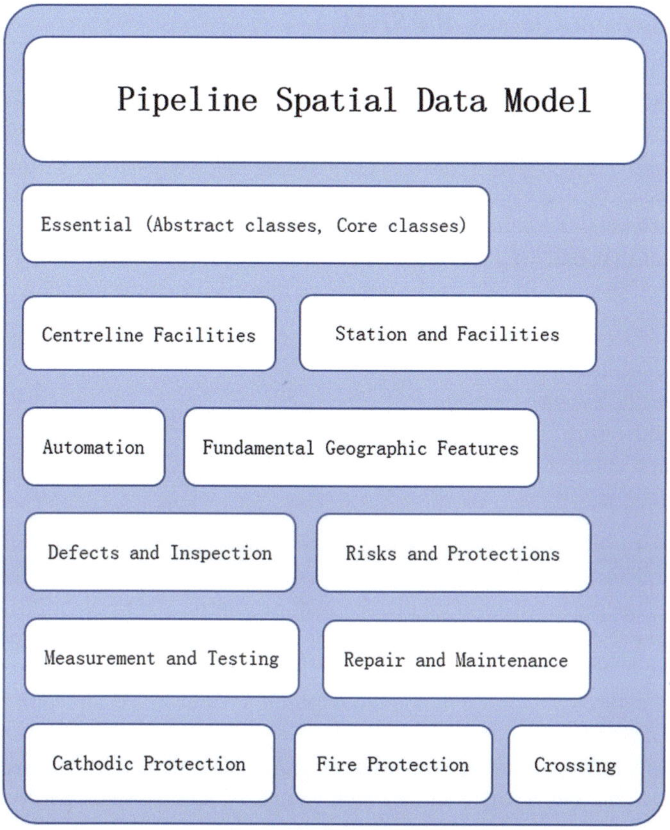

Fig. 3.10 PSDM modules

pipeline companies, there is no standard for how to carry out hierarchical division. Most pipeline companies have some sort of hierarchy of pipeline systems.

In pipeline system, the most common hierarchical unit is LineLoop. In PSDM, LineLoop, composed of several StationSeries, can be further divided into physical LineLoop and logic LineLoop. In general, a continuous StationSeries without branches forms a physical LineLoop, which is usually separated by large equipment or site. Several physical LineLoops form a logical LineLoop. It is worth noting that a logic LineLoop can also contain several sub logic LineLoops, which constitute a relatively complicated LineLoop hierarchical relationship.

Another more common hierarchical relationship is a subsystem. Pipelines often span different administrative or geographic areas. Pipeline within each area, referred to as subsystems, can then be subdivided into more specific subsystems so as to form complex subsystem hierarchical relationship.

PSDM provides modeling support for both hierarchical relationships. See the introduction to PSDM Core Classes for details.

3.4.6 Abstract Classes of PSDM

An Abstract class defines common attributes and behaviors of certain types of pipeline features. The differences of certain types of pipeline features can be distinguished through Subtypes. Abstract Classes do not appear in the Geodatabase as feature classes or object classes. An Entity Class that appears in the Geodatabase inherits attributes and behaviors from an Abstract Class, and unique attributes and behaviors of that Entity Class can be added.

Abstract Classes of PSDM can be organized into the following four classes according to different functions and contents:

(1) Foundation Classes,
(2) Centerline Feature Classes,
(3) Online Feature Classes, and
(4) Offline Feature Classes.

Foundation Classes provide descriptions of basic data of pipeline features, including six classes of PSDMObject, ObjectArchive, NonFacilityObject, FacilityObject, Feature, and FeatureArchive. Feature, a class with geometric attributes in Geodatabase, is the parent class of all classes with geometric attributes in PSDM.

Centerline Feature Classes include four classes of CenterlineObject, CenterlinePolyline, CenterlinePolylineEvent, and CenterlinePoint.

Online Feature Classes are divided into Online Feature Class and Online Feature Class for Offline Feature.

Online Feature Classes include seven classes of OnlineFeature, OnlinePolyline, OnlinePoint, OnlineFacility, OnlinePolylineFacility, OnlinePointFacility, and Fitting.

Online Feature Classes for Offline Feature consists of OnlinePolylineForOfflineFeature and OnlinePointForOfflineFeature.

Offline Feature Classes contain three classes of OfflineFeature, OfflineFacility, and OfflineSiteFacility.

As stated above, PSDM Abstract Classes are also divided into Object Class and Feature Class. See Figs. 3.11 and 3.12 for the object model structure of PSDM Abstract Object Class and Abstract Feature Class.

The detailed lists of PSDM Abstract Classes and descriptions are as follows.

(1) PSDMObject: Its definition and description are shown in Table 3.1.

PSDMObject is the parent class of all of the object classes (indicating features not containing geometric field) in PSDM Abstract Class. The purpose of PSDMObject class is to propagate the EventID attribute. EventID attribute, which must be included in most features or objects in PSDM, represents a globally unique identifier of GUID. The use of GUID as an identifier guarantees the uniqueness of features and objects, even if features and objects are exported and imported into the Geodatabase.

Most of features or objects in PSDM must contain EventID attribute, with the exception of two classes, LineLoopHierarchy and SubSystemHierarchy which are

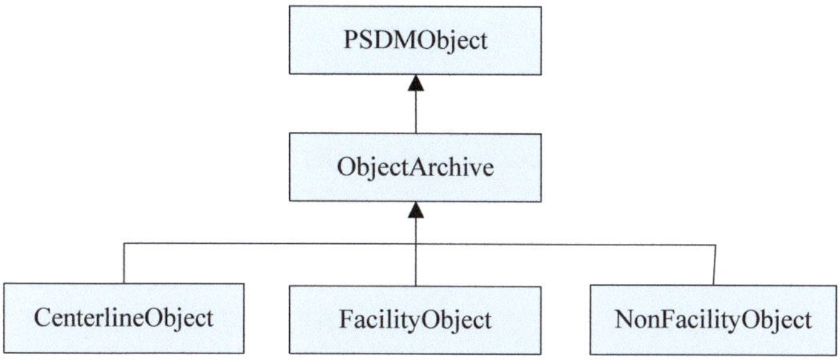

Fig. 3.11 Object model structure of PSDM Abstract Object Class. Re-edited with the permission from Ref. [40]. Esri Image is used herein with permission.

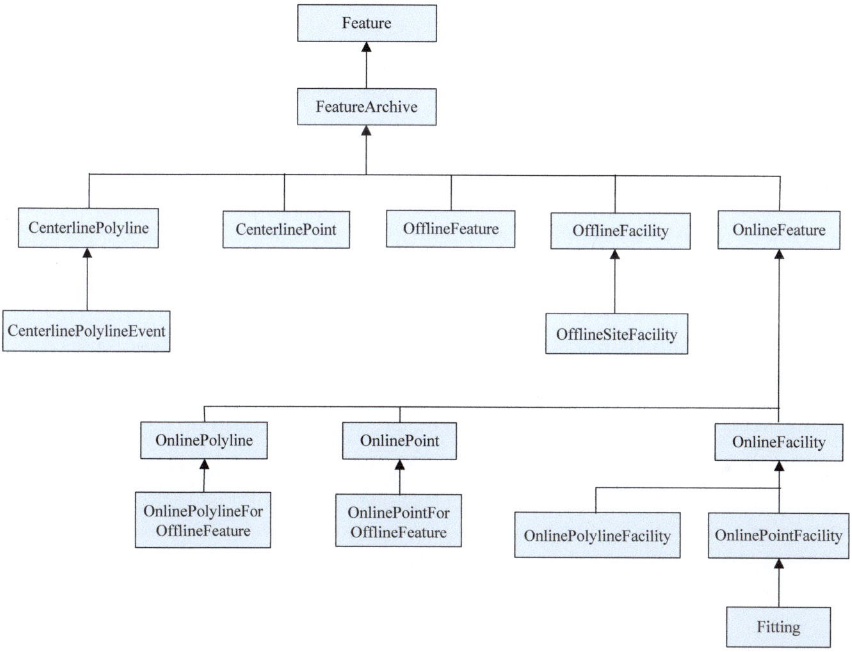

Fig. 3.12 Object model structure of PSDM Abstract Feature Class. Re-edited with the permission from Ref. [40]. Esri Image is used herein with permission.

Table 3.1 Definition and description of PSDMObject

Attribute	Description	Value type
EventID	It represents a globally unique identifier that contains a GUID	String
Geometric type	No	
Inheritance relationship	No	
Relationship with other classes	No	
Subtype	No	

described in the following part. These two classes do not define EventID attribute because the existence and behavior of these two classes are fully described by their parent classes (LineLoop and SubSystem, respectively).

It should be noted that although EventID attribute contains a word "Event", it does not mean that the attribute is only related to the event. In fact, the attribute can be used to identify any event related to the pipeline system (such as leakage) or a feature (such as an oil station, a device in a gas station). The use of "event" is primarily to match the customary title for the linear referencing system.

(2) ObjectArchive: Its definition and description are shown in Table 3.2.

ObjectArchive contains object archive information, including user of creating the object, user's or application's names of the last modifying the object, effective date, ineffective date, and historical state. The information is mainly used for object operational state query and historical data backtracking.

(3) CenterlineObject: Its definition and description are shown in Table 3.3.

The CenterlineObject class provides access to OriginEventID and GroupEventID, of which typical usage is to manage the fact that one online polyline feature crosses two StationSeries or is split, which are used for such hierarchical objects as LineLoop and SubSystem. When using the field OperationalStatus of LineLoop and SubSystem objects, its field value will be propagated to all subobjects of LineLoop or SubSystem objects.

(4) NonFacilityObject: Its definition and description are shown in Table 3.4.

NonFacilityObject is the object that does not have operational meaning or does not exist as a geographic or physical object. NonFacilityObject abstract class provides field Status for nonfacility objects. Status describes the basic status rather than the full operational status.

(5) FacilityObject: Its definition and description are shown in Table 3.5.

FacilityObject Abstract Class provides attributes such as InServiceDate, InstallationDate, and OperationalStatus for objects representing the facility. Subclass objects inherited from this class generally represent objects that have no geometric attributes, such as compressor engines and valve operators. SiteEventID of the class is used to associate the facility with the site to which the facility belongs.

Table 3.2 Definition and description of ObjectArchive

Attribute	Description	Value type
OriginEventID	Feature original identifier. OriginEventID is the same as the feature EventID in creation of a feature. OriginEventID remains the same throughout the life of the feature. OriginEventID is mainly used in the division of polyline features. When a polyline feature is split into multiple subsegments, each subsegment generates a new EventID, but each subsegment also inherits original OriginEventID attributes from the polyline feature and has the same attribute value. Such a mechanism is conducive to maintaining links between features	String
Created By	Creator	String
CreatedDate	Created date	Date
EffectiveFromDate	Object effective date. For a feature created for the first time, the effective date corresponds to the feature's InServiceDate or InstallationDate. Since some objects or features are stored in the database after they are created, they will take effect when they are used, invalidate when uninstalled, and take effect again when they are used again. The attribute value is modified by the next historical state record	Date
EffectiveToDate	Object ineffective date. For a running object, the attribute value is null. Similar to the effective date, the attribute value is modified by the next historical state record	Date
LastModified	The last modified date of the object	Date
Modified By	The last modifier of the object	String
HistoricalState	Object historic status. The data type is the necessary domain of gnHistoricalState of PSDM. This attribute, indicating the historical state of the object, currently supports three states: 0 unknown, 1 current state, and 2 historical state	Long integer
Remarks	Note	String
Geometric type	No	
Inheritance relationship	PSDMObject	
Relationship with other classes	No	
Subtype	No	

Table 3.3 Definition and description of CenterlineObject

Attribute	Description	Value type
GroupEventID	For LineLoop and SubSystem, it is used for organizing all of the subobjects of the two objects, of which GroupEventID value is equal to EventID of the two objects	String
OperationalStatus	OperationalStatus is applied to Centerline and facility features. The data type is the necessary domain of gnOperationalStatus of PSDM	Long integer
Geometric type	NO	
Inheritance relationship	PSDMObject → ObjectArchive	
Relationship with other classes	No	
Subtype	No	

Table 3.4 Definition and description of NonFacilityObject

Attribute	Description	Value type
Status	It describes NonFacilityObject status. The data type is the necessary domain of gnStatus of PSDM	Long integer
Geometric type	No	
Inheritance relationship	PSDMObject → ObjectArchive	
Relationship with other classes	No	
Subtype	No	

Table 3.5 Definition and description of FacilityObject

Attribute	Description	Value type
InServiceDate	The time when a device is put into use. It may be later than the Installation Date	Date
InstallationDate	Installation time	Date
OperationalStatus	Operational status of the object	Long integer
SiteEventID	It specifies the site to which the facility belongs	String
Geometric type	No	
Inheritance relationship	PSDMObject → ObjectArchive	
Relationship with other classes	No	
Subtype	No	

(6) FeatureArchive: Its definition and description are shown in Table 3.6.

FeatureArchive Abstract Class is the parent class of all feature classes in PSDM. The class inherits from Feature Class of Geodatabase model, which has geometric attribute of Shape, so all subclasses inherited from FeatureArchive Abstract Class have geometric attributes. The main purposes of the class are that 1. propagating the EventID attribute; 2. storing object's historical archive information, which is used for object's operational status query and historical data backtracking.

(7) CenterlinePolyline: Its definition and description are shown in Table 3.7.

Table 3.6 Definition and description of FeatureArchive

Attribute	Description	Value type
CreatedBy	Object creator	String
CreatedDate	Created date of the object	
EffectiveFromDate	The object EffectiveFromDate. For a feature created for the first time, the effective date corresponds to the feature's InServiceDate or InstallationDate. Since some objects or features are stored in the database after they are created, they will take effect when they are called, invalidate when uninstalled, and take effect again when they are called again. The attribute value is modified by the next historical state record	Date
EffectiveToDate	The object ineffective date. For a running object, its value is null. Similar to the effective date, the attribute value is modified by the next historical state record	Date
LastModified	The last modified date of the object	Date
ModifiedBy	The last modifier of the object	String
HistoricalState	The object historic status. The data type is the necessary domain of gnHistoricalState of PSDM. This attribute, indicating the historical state of the object, currently supports three states: 0 unknown, 1 current state, and 2 historical state	Long integer
Remarks	Note	String
Geometric type	No	
Inheritance relationship	Feature	
Relationship with other classes	No	
Subtype	No	

Table 3.7 Definition and description of CenterlinePolyline

Attribute	Description	Value type
OperationalStatus	Operational status of the object	Long integer
Geometric type	Polyline feature	
Inheritance relationship	Feature → FeatureArchive	
Relationship with other classes	No	
Subtype	No	

Table 3.8 Definition and description of CenterlinePolylineEvent

Attribute	Description	Value type
StationSeriesEventID	StationSeries to which the object belongs	String
BeginStation	The measure value of the starting point of the object on the StationSeries	Double
EndStation	The measure value of the ending point of the object on the StationSeries	Double
GroupEventID	It has the same meaning as the CenterlineObject class	String
Geometric type	Polyline or as object classes stored in the event table	
Inheritance relationship	Feature → FeatureArchive → CenterlinePolyline	
Relationship with other classes	M: 1—StationSeries	
Subtype	No	

The CenterlinePolyline Abstract Class is the parent class for polyline feature class related to Centerline, providing access to the operational status of the feature.

(8) CenterlinePolylineEvent: Its definition and description are shown in Table 3.8.

This class mainly represents non-facility Centerline feature. Therefore, compared with online polyline facility class, CenterlinePolylineEvent does not have such attributes as InstallationDate and InServiceDate. Subclasses inherited from this class can be implemented as feature classes with geometric attributes, or as object classes stored in event tables. Therefore, the class adds several attributes to support dynamic segmentation including StationSeriesEventID, BeginStation, EndStation, and GroupEventID. A core class of PSDM, SubSystemRange, inherits from this class.

(9) CenterlinePoint: Its definition and description are shown in Table 3.9.

The class represents points that build the Centerline. Currently, only the ControlPoint core class inherits from the class.

(10) OfflineFeature: Its definition and description are shown in Table 3.10.

Table 3.9 Definition and description of CenterlinePoint

Attribute	Description	Value type
GroupEventID	The attribute is used to organize all points that exist in a physical location for purposes such as querying, reporting, and calculating	String
OperationalStatus	The operational status of the object	Long integer
StationSeriesEventID	The StationSeries to which the object belongs	String
Geometric type	Polyline feature	
Inheritance relationship	Feature → FeatureArchive	
Relationship with other classes	M: 1— StationSeries	
Subtype	No	

Table 3.10 Definition and description of OfflineFeature

Attribute	Description	Value type
Status	The object status	Long integer
Geometric type	A point, polyline, or polygon feature without measure value	
Inheritance relationship	Feature → FeatureArchive	
Relationship with other classes	0, 1: M—OnlinePoint 0, 1: M—OnlinePolylineForOfflineFeature	
Subtype	No	

OfflineFeature stores the geometric attributes of point, polyline, and polygon. The positioning system of the object of the class is a geographical reference coordinate system instead of a linear referencing system. OfflineFeature generally refers to objects that assist in the operation and description of the pipeline system, such as geographic data and ancillary buildings.

(11) OfflineFacility: Its definition and description are shown in Table 3.11.

OfflineFacility represents offline facility features that provide access to attributes of InServiceDate, InstallationDate, and OperationalStatus. PSDM's Site core class inherits from this class. The Site class object represents the area feature with boundaries, such as a site, a storage area, and possibly a large building that contains other facilities. The InstallationDate of the Site class refers to the built time.

(12) OfflineSiteFacility: Its definition and description are shown in Table 3.12.

The class represents the facility inside a site. SiteEventID specifies the site to which the facility belongs. Once the site and facility have assigned affiliation, the site becomes a container of facilities, and the facilities within the site can no longer have

Table 3.11 Definition and description of OfflineFacility

Attribute	Description	Value type
InServiceDate	The time when a facility is put into use. It may be later than the InstallationDate	Date
InstallationDate	Installation date	Date
OperationalStatus	Operational status of the object	Long integer
Geometric type	Polygon feature without measure value	
Inheritance relationship	Feature → FeatureArchive	
Relationship with other classes	0, 1: M—OnlinePoint 0, 1: M—OnlinePolylineForOfflineFeature	
Subtype	No	

Table 3.12 Definition and description of OfflineSiteFacility

Attribute	Description	Value type
SiteEventID	It specifies the site to which the facility belongs	String
Geometric type	A point, polyline, or polygon feature without measure value	
Inheritance relationship	Feature → FeatureArchive → OfflineFacility	
Relationship with other classes	M: 1—Site	
Subtype	No	

geometric attributes. Through the affiliation of the site and the facility, all of the facilities' features included in the site can be queried.

(13) OnlineFeature: Its definition and description are shown in Table 3.13.

The class represents features on the Centerline. Online feature can only be point feature or polyline feature, and it must be strictly on the Centerline and use linear referencing for positioning.

Table 3.13 Definition and description of OnlineFeature

Attribute	Description	Value type
StationSeriesEventID	The StationSeries to which the online feature belongs	String
Geometric type	Point, polyline features or as object classes stored in the event table	
Inheritance relationship	Feature → FeatureArchive	
Relationship with other classes	M: 1—StationSeries	
Subtype	No	

(14) OnlinePolyline: Its definition and description are shown in Table 3.14.

The OnlinePolyline online abstract class encapsulates the attributes and behaviors of non-facility online polyline features. Non-facility features refer to features that do not exist in the real world, such as a pressure test on the pipeline, so Status is used to describe the status of non-facility features. Online facilities refer to the physical features that exist in the real world, so OperationalStatus is used to describe the state of the facility features.

(15) OnlinePoint: Its definition and description are shown in Table 3.15.

The class provides access to the status attributes of the non-facility online point features and the measure value attributes along the StationSeries. The location of the online point is determined by StationSeries (specified by the StationSeriesEventID inherited from OnlineFeature) and the specific measure value (Station attribute value).

(16) OnlinePointForOfflineFeature.

Consider the following conditions:

1. It is possible for pipelines to cross certain entities such as rivers and highways. "Crossing" itself is an event and can be modeled as an offline feature, but the starting point and the ending point of the pipeline crossing are closely related to the Centerline of the pipeline.

Table 3.14 Definition and description of OnlinePolyline

Attribute	Description	Value type
BeginStation	The measure value of the starting point of the object on a StationSeries	Double
EndStation	The measure value of the ending point of the object on a StationSeries	Double
GroupEventID	It has the same meaning as the CenterlineObject class	String
Status	The object status	Long integer
Geometric type	Polyline feature	
Inheritance relationship	Feature → FeatureArchive → OnlineFeature	
Relationship with other classes	0, 1: M—OnlinePoint 0, 1: M—OnlinePolyline	
Subtype	No	

Table 3.15 Definition and Description of OnlinePoint

Attribute	Description	Value type
Station	The measure value of the object along the StationSeries	Double
Status	The object status	Long integer
Geometric type	The point feature	
Inheritance relationship	Feature → FeatureArchive → OnlineFeature	
Relationship with other classes	M: 0, 1— OnlinePolyline	
Subtype	No	

2. The milepost can be modeled as an offline facility, but a reference point on the Centerline is required to record the offset location of the milepost to the Centerline.

Offline objects are positioned by geographic coordinates, while the Centerline uses a different positioning system. However, in the above case, it is necessary to locate the offline object based on the Centerline. For this purpose, two classes are designed to maintain the location relationship between offline objects and Centerlines. The two classes are OnlinePointForOfflineFeature and OnlinePolylineForOfflineFeature. The definition and description of OnlinePointForOfflineFeature are shown in Table 3.16.

The offline objects previously referred to include offline features and offline facility objects.

Table 3.16 Definition and description of OnlinePointForOfflineFeature

Attribute	Description	Value type
<OfflineFeature>EventID	It specifies EventID of offline object related to online point features, <OfflineFeature> represents the class name of offline object	String
OffsetAngle	The angle of the line connected by the online point and the offline object and the Centerline	Double
OffsetDistance	The linear distance between online point feature and offline object	Double
Geometric type	Point features or as object classes stored in the event table	
Inheritance relationship	Feature → FeatureArchive → OnlineFeature → OnlinePoint	
Relationship with other classes	M: 1— OfflineFacility M: 1—OfflineFeature	
Subtype	No	

OnlinePointForOfflineFeature describes the connection between an online point feature and an offline object. The online point feature can be the intersection of the Centerline and the offline polyline feature, or it can be the closest online point to an offline point feature, an offline polyline feature, or an offline polygon feature. The class inherits from OnlinePoint and therefore inherits the Station attribute, which indicates the measure value of the online point feature on a StationSeries. The offline object may have no intersection with the Centerline, but has a certain offset from the Centerline. The class defines OffsetAngle and OffsetDistance to describe the offset relationship. Through these attributes, location connection can be generated between an offline object and an online point of the Centerline.

(17) OnlinePolylineForOfflineFeature: Its definition and description are shown in Table 3.17.

OnlinePolylineForOfflineFeature describes the relationship between a segment of a polyline feature on the Centerline and an offline object (usually a polyline feature or a polygon feature). The online polyline feature may be generated by a Centerline crossing some entities, or a Centerline overlapping some polygon features. The class inherits from OnlinePolyline and therefore has BeginStation and EndStation attribute, which indicates the starting and end measure values of the polyline feature

Table 3.17 Definition and description of OnlinePolylineForOfflineFeature

Attribute	Description	Value type
<OfflineFeature>EventID	It specifies EventID of offline object related to online linear features, <OfflineFeature> represents the class name of offline object	String
BeginOffsetAngle	It is the angle of the line (connecting the starting point of online linear feature to the offline object) and the Centerline	Double
BeginOffsetDistance	The linear distance between the starting point of online linear feature and the offline object	Double
EndOffsetAngle	It is the angle of the line (connecting the ending point of online linear feature and the offline object) and the Centerline	Double
EndOffsetDistance	It is the linear distance between the ending point of online linear feature and the offline object	Double
Geometric type	Polyline features or as object classes stored in the event table	
Inheritance relationship	Feature → FeatureArchive → OnlineFeature → OnlinePolyline	
Relationship with other classes	M: 1—OfflineFacility M: 1— OfflineFeature	
Subtype	No	

Table 3.18 Definition and description of OnlineFacility

Attribute	Description	Value type
InServiceDate	The time when the facility is put into use. It may be later than the InstallationDate	Date
InstallationDate	Installation date	Date
OperationalStatus	Operational status of the object	Long integer
SiteEventID	It specifies the site to which the facility belongs	String
Geometric type	Point, polyline, or as object classes stored in the event table	
Inheritance relationship	Feature → FeatureArchive → OnlineFeature	
Relationship with other classes	M: 1—Site	
Subtype	No	

on a StationSeries. It is possible that the offline object has no intersection with the Centerline, but has a certain offset from the Centerline. This is the same for OnlinePointForOfflineFeature. The class also defines OffsetAngle and OffsetDistance to describe the offset relationship. Through these attributes, location connection can be generated between an offline object and an online polyline feature on the Centerline.

(18) OnlineFacility: Its definition and description are shown in Table 3.18.

OnlineFacility encapsulates the attributes and behaviors of online facilities represented by online feature classes. These attributes include InServiceDate, InstallationDate, and OperationalStatus, etc.

(19) OnlinePointFacility: Its definition and description are shown in Table 3.19.

OnlinePointFacility encapsulates the attributes and behaviors of facilities represented by online point features. Station specifies the measure value of the object along the StationSeries.

(20) OnlinePolylineFacility: Its definition and description are shown in Table 3.20.

Table 3.19 Definition and description of OnlinePointFacility

Attribute	Description	Value type
Station	The measure value of the object along a StationSeries	Double
Geometric type	Point features or as object classes stored in the event table	
Inheritance relationship	Feature → FeatureArchive → OnlineFeature → OnlineFacility	
Relationship with other classes	M: 1—Site	
Subtype	No	

Table 3.20 Definition and description of OnlinePolylineFacility

Attribute	Description	Value type
BeginStation	Measure value of the object starting point along StationSeries	Double
EndStation	Measure value of the object ending point along StationSeries	Double
GroupEventID	The same meaning as the CenterlineObject class	String
Geometric type	Polyline features or as object classes stored in the event table	
Inheritance relationship	Feature → FeatureArchive → OnlineFeature → OnlineFacility	
Relationship with other classes	M: 1—Site	
Subtype	No	

OnlinePolylineFacility encapsulates the attributes and behaviors of facilities represented by online polyline features. Different from OnlinePointFacility, OnlinePolylineFacility requires two measure values (BeginStation and EndStation) to determine the online position of the polyline features.

(21) Fitting: Its definition and description are shown in Table 3.21.

Fitting inherits from OnlinePointFacility, but adds the common attributes of fittings, including manufactured date, grade, material, and specification. All features listed as fitting class will inherit from this class.

3.4.7 Core Classes of PSDM

In PSDM, Abstract Classes define common characteristics of the PSDM framework and pipeline feature classes, and Core Classes of PSDM inherit from Abstract Classes. Differing from Abstract Classes, Core Classes are the classes that can be instantiated, that is, specific objects can be created from Core Classes. Core Classes define Centerline features, linear referencing, and other key features of PSDM. Core Classes are divided into Core Object Class and Core Feature Class. Core Object Classes are used to define objects that do not contain geometric attributes, such as the linear referencing system, lineloop, and subsystem. Currently, PSDM Core Object Class includes the following eight classes:

- Company.
- ExternalDocument,

Table 3.21 Definition and description of fitting

Attribute	Description	Value type
DateManufactured	Manufactured date	Date
Grade	Grade	Long integer
InletConnectionType	Inlet connection type	Long integer
InletDiameter	Inlet diameter	Long integer
InletWallThickness	Inlet wall thickness	Long integer
Manufacturer	Manufacturer	Long integer
Material	Material	Long integer
PressureRating	Pressure rating	Long integer
Specification	specification	Long integer
Geometric type	Point features or as object classes stored in the event table	
Inheritance relationship	Feature → FeatureArchive → OnlineFeature → OnlineFacility → OnlinePointFacility	
Relationship with other classes	No	
Subtype	No	

- LineLoop,
- LineLoopHierarchy,
- ReferenceMode,
- Product,
- SubSystem, and
- SubSystemHierarchy.

The object model structure of PSDM Core Object Class is shown in Fig. 3.13.

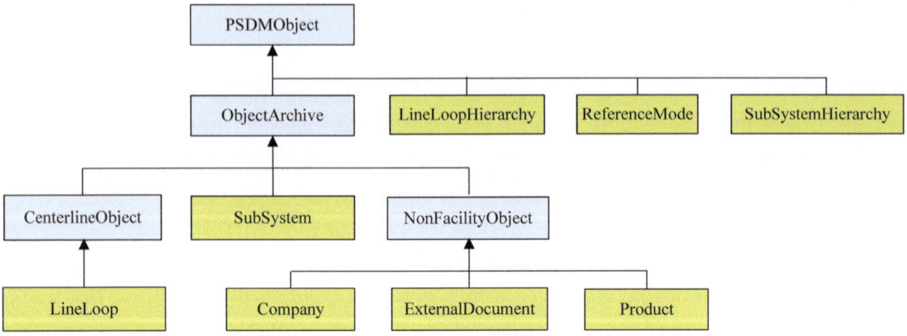

Fig. 3.13 Object model structure of PSDM Core Object Class. Re-edited with the permission from Ref. [40]. Esri Image is used herein with permission. Copyright © 2019 Esri. All rights reserved

Core Feature Classes are used to create pipeline spatial entity features of ControlPoint and Stationseries. At present, PSDM Core Feature Class consists of the following four classes:

- ControlPoint,
- Site,
- StationSeries, and
- SubSystemRange.

The object model structure of PSDM Core Feature Class is shown in Fig. 3.14.
The detailed lists of PSDM Core Classes and descriptions are as follows.

(1) Company: Its definition and description are shown in Table 3.22.

Company is used to store the information of companies involved in pipeline system activities (includes operation management, repair, maintenance, and supply).

(2) Product: Its definition and description are shown in Table 3.23.

Product is used to store the type information of products transmitted by the pipelines. A single product (such as natural gas) can be transmitted through multiple pipelines, while a single pipeline can transmit different products (such as crude oil) in sequence at the same time. Therefore, many-to-many relationships are defined between Product and Pipeline.

(3) ExternalDocument: Its definition and description are shown in Table 3.24.

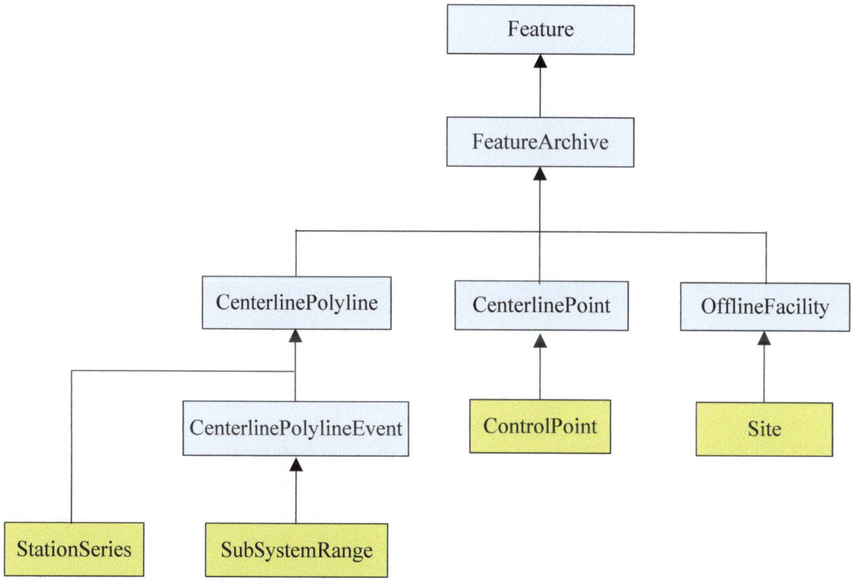

Fig. 3.14 Object model structure of PSDM Core Feature Class. Re-edited with the permission from Ref. [40]. Esri Image is used herein with permission. Copyright © 2019 Esri. All rights reserved

Table 3.22 Definition and description of company

Attribute	Description	Value type
CompanyLabel	Company labels, abbreviations, or other titles	String
CompanyName	Company name	String
CompanyType	Service type provided by the company, such as pipeline transmission and pipeline construction	Long integer
Geometric type	No	
Inheritance relationship	PSDMObject → Object Archive → NonFacilityObject	
Relationship with other classes	No	
Subtype	No	

Table 3.23 Definition and description of product

Attribute	Description	Value type
Product	Product type, e.g., crude oil and natural gas	Integer
Geometric type	No	
Inheritance relationship	PSDMObject → Object Archive → NonFacilityObject	
Relationship with other classes	M: N—LineLoop	
Subtype	No	

Table 3.24 Definition and description of ExternalDocument

Attribute	Description	Value type
DocumentDescription	Description of the external document	String
DocumentType	Type of the external document	Long integer
FilePath	File path of the external document	String
FileName	Filename of the external document	String
HyperLink	Hyperlink of the external document	String
Geometric type	No	
Inheritance relationship	PSDMObject → Object Archive → NonFacilityObject	
Relationship with other classes	No	
Subtype	No	

Data of certain features or events, for some reasons, may be stored in the form of external documents in a physical hard disk, network folder, document management system, etc., instead of a pipeline database. The external document class stores the information of an external document related to a feature, including location, type, and name.

(4) LineLoop: Its definition and description are shown in Table 3.25.
(5) LineLoopHierarchy: Its definition and description are shown in Table 3.26.

LineLoop and LineLoopHierarchy are both used to model Lineloop hierarchy. LineLoop consists of a series of interconnected StationSeries. One LineLoop can have several parent LineLoops, and can also contain multiple child LineLoops. LineLoopHierarchy is used to describe the relationship between the parent LineLoop and the child LineLoop. Its ParentLineLoopEventID indicates the EventID of the parent LineLoop, and ChildLineLoopEventID specifies the EventID of a child LineLoop. If there are multiple child LineLoops in a parent LineLoop, the corresponding number of records need to be stored in the LineLoopHierarchy table.

Figure 3.15 is an example showing that #1 long-distance pipeline system operated by a pipeline company contains two main lines (Line #1 and Line #6) and one branch

Table 3.25 Definition and description of Lineloop

Attribute	Description	Value type
LineName	Line name	String
LineType	Line type, e.g., "gathering and transportation" and "transmission"	Long integer
Geometric type		
Inheritance relationship	PSDMObject → Object Archive → CenterlineObject	
Relationship with other classes	1: M—StationSeries 1: M—LineLoopHierarchy M: N—Product	
Subtype		

Table 3.26 Definition and description of LineLoopHierarchy

Attribute	Description	Value type
ParentLineLoopEventID	EventID of parent LineLoop	String
ChildLineLoopEventID	EventID of child LineLoop	String
Geometric type	No	
Inheritance relationship	PSDMObject	
Relationship with other classes	M: 1—LineLoop	
Subtype	No	

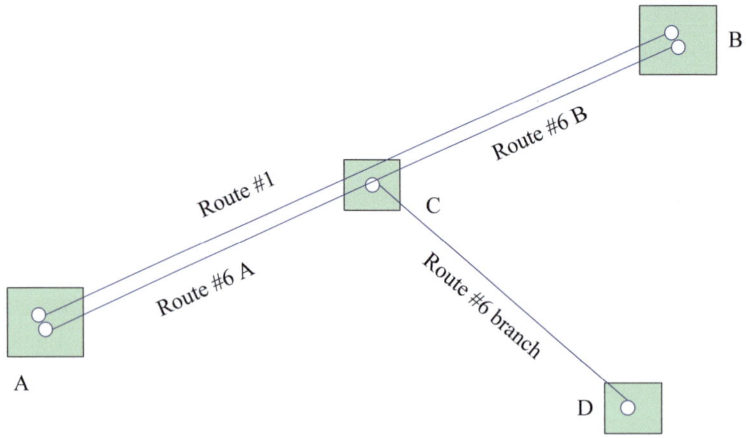

Fig. 3.15 LineLoop hierarchy diagram of long-distance pipeline System #1 of a pipeline company. Re-edited with the permission from Ref. [40]. Esri Image is used herein with permission. Copyright © 2019 Esri. All rights reserved

line (Branch Line #6). The long-distance pipeline system consists of six LineLoops, of which the long-distance pipeline system #1 and the main Line #1 are logical LineLoops. Long-distance pipeline system #1 consists of three child LineLoops: Line #1, Line #6 and Branch Line #6. Main Line #6 consists of two child LineLoops: Line #6 A and Line #6 B.

The records of LineLoop hierarchical relationship of the pipeline system in LineLoopHierarchy are shown in Table 3.27.

(6) ReferenceMode: Its definition and description are shown in Table 3.28.

The class is used to define the linear referencing for ControlPoints and LineLoops, including linear referencing methods and linear referencing units. One linear referencing method must be defined for each LineLoop and ControlPoint, and the LineLoop and its ControlPoints must adopt the same linear referencing method. In PSDM, all online features are positioned based on a linear referencing method.

(7) SubSystem: Its definition and description are shown in Table 3.29.

Table 3.27 LineLoop hierarchical relationship of long-distance pipeline system #1

ParentLineLoopEventID	ChildLineLoopEventID
Long-distance Pipeline System #1	Line #1
Long-distance Pipeline System #1	Line #6
Long-distance Pipeline System #1	Line #6 branch
Line #6	Line #6 A
Line #6	Line #6 B

Table 3.28 Definition and description of ReferenceMode

Attribute	Description	Value type
RefModeMethod	Linear referencing methods, e.g., "milepost reference method" or "3D projected method"	Long integer
RefModeUnits	Linear referencing units	Long integer
Geometric type	No	
Inheritance relationship	PSDMObject	
Relationship with other classes	No	
Subtype	No	

Table 3.29 Definition and description of SubSystem

Attribute	Description	Value type
SubSystemName	SubSystem name	String
Geometric type	No	
Inheritance relationship	PSDMObject → ObjectArchive	
Relationship with other classes	1: M—SubSystemHierarchy 1: M—SubSystemRange M: N—Site	
Subtype	No	

(8) SubSystemHierarchy: Its definition and description are shown in Table 3.30

Subsystem is a logical unit derived from the division of pipelines on a certain basis such as administrative divisions, geographical phenomena, and commercial and economic factors. Subsystem and LineLoop have the same hierarchical structure, that is, there can be multiple parent systems in a subsystem, and a subsystem can also contain multiple subsystems. For example, a long-distance pipeline spanning two provinces forms a subsystem in each province, while the pipeline spanning multiple counties also forms a subsystem in each county. The province subsystem and the

Table 3.30 Definition and description of SubSystemHierarchy

Attribute	Description	Value type
ParentSubSystemEventID	Parent system EventID	String
ChildSubSystemEventID	Child subsystem EventID	String
Geometric type	No	
Inheritance relationship	PSDMObject	
Relationship with other classes	M: 1—SubSystem	
Subtype	No	

county subsystem form a parent-child relationship. It is possible to obtain subsystems by separating successive LineLoop or StationSeries.

SubSystem and SubSystemHierarchy are used together to model the subsystem hierarchy. SubSystemHierarchy is used to store the relationship between the parent system and the child system. Its ParentSubSystemEventID indicates the parent system EventID, while ChildSubSystemEventID specifies a child subsystem EventID of the system. If there are multiple child subsystems in a parent system, the corresponding number of records needs to be stored in the SubSystemHierarchy table.

Figure 3.16 shows a Long-distance Pipeline System #1 spanning district E and district F to form three subsystems: the Long-distance Pipeline Administrative System #1, the District E Subsystem, and the District F Subsystem. The two subsystems are the child systems of Long-distance Pipeline Administrative System #1. The District E Subsystem consists of five pipeline segments: SSR-1, SSR-2, SSR-6, SSR-7, and SSR-9. The District F Subsystem includes four pipeline segments: SSR-3, SSR-4, SSR-5, and SSR-8.

The records of subsystem hierarchical relationships of the pipeline system in SubSystemHierarchy are shown in Table 3.31.

The range of the pipeline corresponding to the subsystem is specified by SubSystemRange. See SubSystemRange for details.

PSDM Core Feature Class is introduced in the following part.

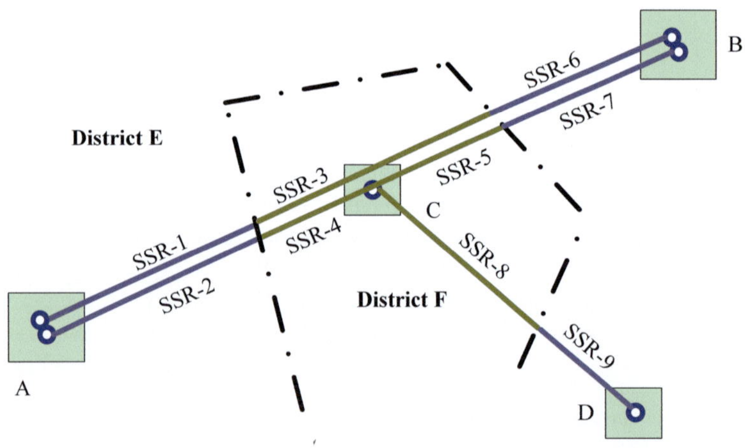

Fig. 3.16 Subsystem hierarchy diagram of the long-distance pipeline system #1 of a pipeline company. Re-edited with the permission from Ref. [40]. Esri Image is used herein with permission.

Table 3.31 Subsystem hierarchical relationship of long-distance pipeline system #1

ParentSubSystemEventID	ChildSubSystemEventID
Long-distance Pipeline System #1	District E Subsystem
Long-distance Pipeline System #1	District F Subsystem

Table 3.32 Definition and description of ControlPoint

Attribute	Description	Value type
ControlPointAngle	The angle between the line formed by connecting the ControlPoint and the previous ControlPoint, and the previous pipeline segment (generally consisting of the previous two ControlPoints of the ControlPoint)	Double
ControlPointType	ControlPoint type, e.g., inflection point, pile, crossing point	Integer
PIDirection	The offset direction of the ControlPoint to the previous pipeline segment (generally consisting of the previous two ControlPoints of the ControlPoint), e.g., "Right side", "Left side", and "No"	Integer
StationValue	Known measure values of ControlPoints along a StationSeries	Double
ReferenceModeEventID	Linear referencing method for ControlPoint	String
Geometric type	Point feature	
Inheritance relationship	Feature → FeatureArchive → CenterlinePoint	
Relationship with other classes	M: 1—StationSeries	
Subtype	No	

(1) ControlPoint: Its definition and description are shown in Table 3.32.

The ControlPoint feature class is the foundation of PSDM. ControlPoints form the Centerline. A control point is the point of the known geographical reference coordinates and the measure value along the Centerline. The series of ControlPoints connected to each other constitute the basic constituent unit of the pipeline Centerline—StationSeries. A StationSeries consists of at least two ControlPoints. A StationSeries and its ControlPoints must share a consistent linear referencing system.

(2) StationSeries: Its definition and description are shown in Table 3.33.

LineLoop forms the Centerline of the pipeline segment. StationSeries of a pipeline can be equipped with different linear referencing modes. Even if it is one StationSeries, different linear referencing modes can be adopted. However, all online features are positioned based on the current referencing mode of the StationSeries.

(3) Site: Its definition and description are shown in Table 3.34.

The class is used to describe features in the pipeline system that has polygon geometric attributes, such as pump stations, meter stations, compressor stations, and temporary work areas.

Site and online or offline facilities form a one-to-many relationship, meaning, a site may contain multiple online or offline facilities.

Table 3.33 Definition and description of StationSeries

Attribute	Description	Value type
SeriesName	StationSeries name	String
BeginStation	The measure value of the StationSeries starting point	Double
EndStation	The measure value of the StationSeries ending point	Double
FromSeriesEventID	A StationSeries may originate from the connection of the previous StationSeries. FromSeriesEventID indicates the EventID of the previous StationSeries connected to the current StationSeries	String
FromConnectionStationValue	The measure value on the current StationSeries of the connection point formed by the current StationSeries and the previous StationSeries	Double
ToSeriesEventID	A StationSeries may originate from the connection of the next StationSeries. ToSeriesEventID indicates the EventID of the next StationSeries connected to the current StationSeries	String
ToConnectionStationValue	The measure value on the current StationSeries of the connection point formed by the current StationSeries and the next StationSeries	Double
LineLoopEventID	It indicates the LineLoop to which the StationSeries belongs. Each StationSeries must belong to a LineLoop	String
ReferenceModeEventID	Linear referencing method for StationSeries	String
Geometric type	Linear feature	
Inheritance relationship	Feature → FeatureArchive → CenterlinePolyline	
Relationship with other classes	1: M—OnlineFeature	
Subtype	No	

Table 3.34 Definition and description of site

Attribute	Description	Value type
SiteName	Site name	String
SiteType	Site type	Long integer
Geometric type	Surface feature	
Inheritance relationship	Feature → FeatureArchive → OfflineFacility	
Relationship with other classes	0, 1: M—OnlineFacility 0, 1: M—OfflineFacility M: N—SubSystem	
Subtype	No	

Table 3.35 Definition and description of SubSystemRange

Attribute	Description	Value type
SubSystemEventID	Indicates the Subsystem to which the SubSystemRange belongs	String
Geometric type	Polyline feature	
Inheritance relationship	Feature → FeatureArchive → CenterlinePolylineEvent	
Relationship with other classes	M: 1—StationSeries	
Subtype	No	

Table 3.36 Relationship of subsystem and subsystem range of long-distance pipeline system #1

SubSystemEventID	SubSystemRange
District E Subsystem	SSR-1
District E Subsystem	SSR-2
District E Subsystem	SSR-6
District E Subsystem	SSR-7
District E Subsystem	SSR-9
District F Subsystem	SSR-3
District F Subsystem	SSR-4
District F Subsystem	SSR-5
District F Subsystem	SSR-8

(4) SubSystemRange: Its definition and description are shown in Table 3.35.

The class is used to define the subsystem range. It inherits from CenterlinePolylineEvent and is a polyline feature with measurement values. BeginStation and EndStation define the range of the subsystem on a StationSeries, which is specified by StationSeriesEventID.

Subsystem and subsystem range constitute a one-to-many relationship, meaning, a subsystem can contain multiple subsystem ranges. For example, in Fig. 3.16, the relationships between the subsystem and the subsystem range are shown in Table 3.36.

3.4.8 Entity Classes and Entity Modules of PSDM

PSDM Entity Classes model specific facilities and events associated with the pipeline and marks the beginning of applying the PSDM data model. All Entity Classes inherit from Abstract Classes and have all of the behaviors and attributes of the inherited Abstract Classes. On this basis, Entity Classes can add additional attributes and behaviors to customize entity objects. The closure is one example.

Closure inherits from Fitting and is an accessory class. Attributes, such as DateManufactured inherited by Closure from Fitting, describes general attributes such as a Closure material. Fitting inherits from OnlinePointFacility, so Closure is an online point facility feature. Closure inherits such attributes as Station from OnlinePoint-Facility through which linear referencing positioning can be carried out on a Closure object. OnlinePointFacility inherits from OnlineFacility, from which Closure inherits such attributes as InstallationDate. OnlineFacility inherits from OnlineFeature, so it is a feature class with geometric fields. Attributes such as StationSeriesEventID, inherited by Closure from OnlineFeature, indicates the LineLoop to which Closure object belongs. OnlineFeature inherits from FeatureArchive, while Closure inherits such attributes as CreatedDate from FeatureArchive; FeatureArchive inherits from Feature, and such attributes as Shape, inherited by Closure from Feature, indicate geometric attributes information of Closure objects. Based on these inherited attributes, Closure adds ClosureType attribute to describe the type of Closure. The complete attribute definition of Closure is shown in Table 3.37.

Table 3.37 Attribute definition of closure

Closure and its parent class	Closure custom attributes and inherited attributes from its parent class
Feature	Shape
FeatureArchive	CreatedBy
	CreatedDate
	EffectiveFromDate
	EffectiveToDate
	LastModified
	ModifiedBy
	HistoricalState
OnlineFeature	StationSeriesEventID
OnlineFacility	InServiceDate
	InstallationDate
	OperationalStatus
	SiteEventID
OnlinePointFacility	Station
Fitting	DateManufactured
	Grade
	InletConnectionType
	InletDiameter
	InletWallThickness
	Manufacturer
	Material
	PressureRating
	Specification
Closure	ClosureType

It is because of the inheritance customization mechanism that PSDM empowers the extensibility. Pipeline companies can extend PSDM to meet their needs.

Based on the data demand analysis of the long-distance oil and gas pipeline, this book divided the Entity classes of PSDM into the following 11 modules:

- Centreline Facilities module,
- Station and Facilities module,
- Automation module,
- Measurement and Testing module,
- Crossing module,
- Defects and Inspection module,
- Cathodic Protection module,
- Risks and Protections module,
- Fire Protection module,
- Repair and Maintenance module, and
- Fundamental Geographic Features module.

Detailed classes listed in each module are as follows:

(1) Centreline Facilities module

Centreline Facilities module models facilities, devices, instruments, and meters located on the pipeline Centerline. Entity classes in this module are listed in Table 3.38.

(2) Stations and Facilities module

Entity classes in this module are listed in Table 3.39.

(3) Automation module

Automation module models SCADA and related automatic instruments. Entity classes in this module are listed in Table 3.40.

(4) Measurement and Testing module

This module models elements that are closely related to pipeline operations (investigation, testing, analysis, etc.). Entity classes in this module are listed in Table 3.41.

(5) Crossing module

This module models the various crossing engineering of the pipeline as well as the interaction between pipelines and external physical features. Entity classes in this module are listed in Table 3.42.

(6) Defects and Inspection module

This module models a variety of defects, inspections, and results. Entity classes in this module are listed in Table 3.43.

(7) Cathodic Protection module

Table 3.38 Centreline facilities classes

Class name	Parent class	Subtype	Description
Casing	OnlinePolylineFacility		
Coating	OnlinePolylineFacility		
Elbow	Fitting		
Meter	Fitting		
PipeJoinMethod	OnlinePointFacility	1—Weld	
		2—Coupling	
		3—Flange	
		4—Screw	
PipeSegment	OnlinePolylineFacility	1—Pipe	
		2—Bend	
		3—Transition	
Reducer	Fitting		
Tap	OnlinePointFacility		
Tee	Fitting		
Valve	OnlinePointFacility	1—Angle valve	
		2—Ball valve	
		3—Block valve	
		4—Check valve	
		5—Control valve	
		6—Gate valve	
		7—Plug valve	
Vessel	OnlinePointFacility		

Entity classes in this module are listed in Table 3.44.

(8) Risks and Protections module

Entity classes in this module are listed in Table 3.45.

(9) Fire Protection module

Entity classes in this module are listed in Table 3.46.

(10) Repair and Maintenance module

Entity classes in this module are listed in Table 3.47.

(11) Fundamental Geographic Features module

Entity classes in this module are listed in Table 3.48.

Table 3.39 Station and Facilities classes

Class name	Parent class	Subtype	Description
Station	OnlinePointFacility/ OfflineFacility	First station	On a small scale map, Station class represents OnlinePointFacility. On a large scale map, Station class is regards as OfflineFacility
		Terminal station	
		Pressure regulating station	
		Compressor station	
		Meter station	
		Distribution station	
		Pigging station	
		Block valve chamber	
Compressor	OfflineSiteFacility		
Pump	OfflineSiteFacility		
OilTank	OfflineSiteFacility		
HeatingFurnace	OfflineSiteFacility		
BurstingDisc	OfflineSiteFacility		
Instrument	OfflineSiteFacility	0—Unknown	
		1—Flow control valve	
		2—Flow computer	
		3—Gas chromatograph	
		4—Gas sampler	
		5—Level controller	
		6—Level indicator	
		7—Liquid sampler	
		8—Pressure controller	
		9—Pressure gauge	
		10—Valve position indicator	
		11—ER probe	
InstrumentParameter	PSDMObject	Same as instrument	
StationValve	OfflineSiteFacility	1—Angle valve	
		2—Ball valve	
		3—Block valve	
		4—Check Valve	
		5—Control valve	
		6—Gate valve	

Table 3.40 Automation classes

Class name	Parent class	Subtype	Description
ControlCenter	OfflineFacility	1—Main control center	
		2—Standby control center	
		3—Station control center	
ComputerServer	OfflineSiteFacility	1—Real-time data server	
		2—Historical data server	
		3—Application server	
		4—Communication server	
WorkStation	OfflineSiteFacility	1—Operator workstation	
		2—Engineer workstation	
Router	OfflineSiteFacility		
PLC	OfflineSiteFacility		Programable logic controler
RTU	OfflineSiteFacility		Remote terminal unit
Transducer	OfflineSiteFacility	1—Pressure transducer	
		2—Temperature transducer	
		3—Flow transducer	
		4—Level transducer	
		5—Water content analyzer	
		6—Densitometer	
		7—Calorific value analyzer	
Actuator	OfflineSiteFacility		
CommunicationLine	OfflineFeature		

3.4.9 PSDM Domain Design

When describing relevant features and events of the pipeline system, often, common values between different pipeline companies can be found, such as valves. All pipeline companies' valves have two states: "ON" and "OFF". The PSDM domain classifies these common values. Each category corresponds to a list each containing several domain values. Each domain value corresponds to a long integer value.

Table 3.41 Measurement and testing classes

Class name	Parent class	Subtype	Description
FieldNote	OfflineFeature	1—Cultural note	
		2—Environmental note	
		3—Facility note	
		4—Geopolitical note	
		5—Hydrology note	
		6—Line crossing note	
		7—Operations note	
		8—Routing note	
		9—Transportation not	
FieldNoteLocation	OnlinePointForOfflineFeature		
Marker	OfflineFeature	1—Milepost	
		2—Aerial marker	
		3—Monument	
		4—Survey point	
		5—PIG signal	
MarkerLocation	OnlinePointForOfflineFeature		
OperatingPressure	OnlinePolyline		
PressureTest	OnlinePolyline		
OnlinePressureMeasurement	OnlinePoint		
OnlineTemepratureMeasurement	OnlinePoint		
OnlineFlowMeasurement	OnlinePoint		
OnlineLevelMeasurement	OnlinePoint		
ElevationPoint	OnlinePoint		
RightOfWay	OnlinePolyline		

Table 3.42 Crossing classes

Class name	Parent class	Subtype	Description
Structure	NonFacilityObject		
NearestPointToLine	OfflineFeature		The location of the nearest point on a structure to a line
StructureLocation	OnlinePointForOfflineFeature		
StructureOutline	OfflineFeature		
LineCrossing	OfflineFeature	1—Geographical	
		2—Utility	
		3—Transportation	
LineCrossingEasement	OnlinePolylineForOfflineFeature		
LineCrossingLocation	OnlinePointForOfflineFeature		

(continued)

Table 3.42 (continued)

Class name	Parent class	Subtype	Description
Encroachment	OnlinePolyline		A segment that is overlapped by ground objects
OverheadPipe	OnlinePolyline		

Table 3.43 Defects and inspection classes

Class name	Parent class	Subtype	Description
Corrosion	OnlinePoint	1—External corrosion	
		2—Internal corrosion	
Dent	OnlinePoint		
Crack	OnlinePoint/OnlinePolyline		
Damage	OnlinePoint		
Defect	OnlinePoint	1—Metal defect	
		2—Welding defect	
		3—Coating defect	
Deformation	OnlinePoint/OnlinePolyline		
InlineInspection	OnlinePolyline	MFL	Magnetic flux leakage testing
		UT	Ultrasonic testing
		RFEC	Remote field eddy current testing
ExternalInspection	OnlinePoint	PCM	pipeline current mapper
		Pearson	
		DCVG	Direct current voltage gradient
		CIPS	Close interval potential method
		P/S	Standard pipe/sub-ground voltage detection technology
		CGDT	Current gradient detection technology
		Aerial survey	
		Excavation	
InspectionRange	OnlinePolyline		
Leak	OnlinePoint		

Table 3.44 Cathodic protection classes

Class name	Parent class	Subtype	Description
CPAnode	OfflineFacility		
CPBond	OfflineFacility		
CPCable	OfflineFacility		
CPGroundBed	OfflineFacility		
CPCoupon	OfflineFacility		
CPLocation	OnlinePointForOfflineFeature	1—CPRectifier	
		2—CPGroundBed	
		3—CPAnode	
		4—CPBond	
		5—CPTestStation	
		6—CPCable	
		7—CPCoupon	
CPRectifier	OfflineFacility		
CPTestStation	OfflineFacility		

Table 3.45 Risks and protections classes

Class name	Parent class	Subtype	Description
UnfavorableGeology	OfflineFeature	1—Landslide	
		2—Debris flow	
		3—Collapse	
		4—Land subsidence	
		5—Goaf	
		6—Others	
FaultZone	OfflineFeature		
HighConsequenceArea	OfflineFeature		
HCARange	OnlinePolyline		
RiskAnalysis	OnlinePolyline		
Protection	OfflineFeature	1—Retaining wall	
		2—Bank protection	
		3—Bottom protection	
		4—Ecological protection	
		5—Slope protection	
		6—Slope consolidation	

Table 3.46 Fire protection classes

Class name	Parent class	Subtype	Description
FireStation	OfflineFeature		
FireEngine	OfflineFeature		
CommunicationCommandCenter	OfflineFeature		
FireEquipment	OfflineFeature	1—Fire hydrant	
		2—Fire extinguisher	
		3—Fire detector	
		4—Alarm	
		5—Fire monitor	
		6—Fire hose	
FirePumpRoom	OfflineFeature		
FireControlRoom	OfflineFeature		

Table 3.47 Repair and maintenance classes

Class name	Parent class	Subtype	Description
Maintenance	OnlinePoint	1—Coating maintenance	
		2—Clamp maintenance	
		3—Composite material maintenance	
		4—Pipeline reinforcement repair	
		5—Welding	
Excavation	OnlinePoint		
StrayCurrent	OfflineFeature		
Sleeve	OnlinePolyline		
ExposedPipe	OnlinePolyline		
CondensingTube	OnlinePolyline		
RepairStation	OfflineFeature		

Different domain values have different meanings. Most domains have a field value: "0", which means "Unknown". A domain usually corresponds to an attribute of a class (including Object Class, Core classes, and Entity Classes) in PSDM, while the field value provides a possible attribute value for the attribute. The domain name is prefixed with two letters to indicate the category type. Different letters represent the following categories, respectively:

- gn: generic domain,
- cp: cathodic protection domain,
- mt: measurement and testing domain,
- fc: facility feature domain,
- cl: centerline feature domain,

Table 3.48 Fundamental geographic features classes

Class name	Parent class	Subtype	Description
District	OfflineFeature		
Road	OfflineFeature	1—Highway	
		2—NationalWay	
		3—ProvincialWay	
		4—Contry road	
Railway	OfflineFeature		
River	OfflineFeature		
City	OfflineFeature		
Lake	OfflineFeature		
DEM	OfflineFeature		Digital elevation model along pipeline
RSImage	OfflineFeature		Remote sensing images along pipeline
NatureReserve	OfflineFeature		
FaultZone	OfflineFeature		
BreakdownService	OfflineFeature		
ManagementOffice	OfflineFeature		
LivingArea	OfflineFeature		
Mountain	OfflineFeature		

- in: inspection domain,
- au: automation domain,
- rp: risks and protections domain,
- fe: fundamental geographic features domain,
- rm: Repair and Maintenance domain, and
- fp: fire protection domain.

Due to limited space, only generic domains are introduced in the following section.

(1) gnOperationalStatus describes the states of a feature with a period of running time and is maiCenterline and facility features. Its field values and meanings are as follows:

- 0 = Unknown;
- 1 = Conceptual, conceived state;
- 2 = Proposed, proposed state;
- 4 = Active, active state;
- 8 = Inactive, inactive state;
- 16 = Idle, idle state;
- 32 = Abandoned, abandoned state; and
- 64 = Removed, removed state.

(2) gnStatus describes non-facility feature states. Its field values and meanings are as follows:

- 0 = Unknown;
- 1 = Conceptual, conceived state;
- 2 = Proposed, proposed state;
- 4 = Active, active state; and
- 8 = Inactive, inactive state.

(3) gnRefModeBasis describes linear referencing method. Its filed values and meanings are as follows:

- 0 = Unknown;
- 1 = MilePost, milepost referencing method;
- 2 = 3D Projected, 3D projected method;
- 3 = 3D Slack Chain, 3D slack chain method;
- 4 = 3D Geoid, 3D geoid method; and
- 5 = 2D Projected, 2D projected method.

(4) gnRefModeUnits describes the unit adopted in linear referencing. Its field values and meanings are as follows:

- 0 = Unknown;
- 1 = Meter, meter;
- 2 = Foot, foot; and
- 3 = Kilometer, kilometer.

(5) gnHistoricalState describes the historical states of an online feature. Its field values and meanings are as follows:

- 0 = Unknown;
- 1 = Current, current state; and
- 2 = Historical, historical state.

(6) gnAngle represents an angle. Its field value, ranging from 0 to 360, has 2 predefined filed values. Its filed values and meanings are as follows:

- MinValue = 0.00,
- MaxValue = 360.00.

(7) gnYesNo indicates a value with two states ("YES" or "NO"). Its field value and meaning are as follows:

- 0 = Unknown;
- 1 = Yes, yes; and
- 2 = No, no.

(8) gnRequiresGeometry indicates whether an object of a feature class allows the geometric attribute value to be null. Its field value and meaning are as follows:

- 0 = Unknown;

- 1 = Must Have, must have geometric attribute;
- 2 = Must Not Have, must not have geometric attribute; and
- 3 = Optional, optional.

3.4.10 PSDM' Support for Pipeline Real-Time Data

Real-time data of the pipeline system, including pipeline temperature, flow, and pump station inlet and outlet pressure, are of vital importance to the operation management of the pipeline. Pipeline SCADA systems, generally installed in today's pipelines, provide pipeline real-time data. Pipeline spatial and attribute information are taken as management goals for most of the current pipeline GIS systems, and pipeline real-time data is less involved. In the PSDM design stage, this author has taken the pipeline real-time data into consideration and incorporated the pipeline real-time data into the data organization of PSDM. On that basis, by integrating the pipeline GIS system and the SCADA system, it is possible to realize real-time monitoring based on the pipeline GIS system.

In PSDM, not all features have real-time data. Real-time data contained in features are different in numbers and types, that is, real-time data does not have common characteristics. It should exist as an attribute of a specific feature that contains real-time data. Therefore, Abstract Class and Core Class of PSDM do not contain real-time data attribute. Real-time data only exists in Entity Classes.

In PSDM, real-time data of the pipeline is divided into the following two types:

(1) Real-time states and readings of certain installed inherent pipeline facilities, equipment, and instruments and apparatus.
This type of Entity Class contains real-time data including oil transportation station (inlet and outlet oil temperature, oil pressure, and flow), pump unit (inlet and outlet oil temperature, oil pressure, flow, bearing temperature, and motor current), oil tank (oil temperature, liquid level, and oil storage), various types of instruments (reading), and valve (opening degree, state). The attribute field of real-time data in this type of Entity Class is represented by real-time data class named plus CP, such as the pressure real-time data field attribute named PressureCP. Here is an entity class that contains real-time data—Meter Entity Classes. It is explained in Table 3.49.
(2) Real-time readings of temporarily installed instruments and apparatuses measure instantaneous working conditions of the pipeline operation. This includes real-time readings of pressure gauges, flow meters, and thermometers all of which are temporarily installed at a certain point in the pipeline for testing and analysis. These temporary measurements are designed as Online Feature Classes, inherited from OnlinePointFacility, and include attributes such as measurements, real-time readings, reading time, reading frequency, instrument models, and instrument types. Currently, PSDM has the following four temporary measurement classes (Table 3.50).

Table 3.49 Attribute definition of meter

Meter and its parent classes	Meter custom attributes and inherited attributes from its parent classes
Feature	Shape
FeatureArchive	CreatedBy
	CreatedDate
	EffectiveFromDate
	EffectiveToDate
	LastModified
	ModifiedBy
	HistoricalState
OnlineFeature	StationSeriesEventID
OnlineFacility	InServiceDate
	InstallationDate
	OperationalStatus
	SiteEventID
OnlinePointFacility	Station
Fitting	DateManufactured
	Grade
	InletConnectionType
	InletDiameter
	InletWallThickness
	Manufacturer
	Material
	PressureRating
	Specification
Meter	MeterFunction
	MeterName
	MeterNumber
	MeterType
	SerialNumber
	ReadingUnits
	ReadingValueCP

Table 3.50 Four temporary measurement classes

Class name	Parent class	Subtype	Description
OnlinePressureMeasurement	OnlinePoint		
OnlineTemepratureMeasurement	OnlinePoint		
OnlineFlowMeasurement	OnlinePoint		
OnlineLevelMeasurement	OnlinePoint		

These classes belong to measurement and testing modules. According to the temporary measurement class, it is possible to integrate instantaneous working conditions such as temperature and pressure at any point on the pipeline, which provides more possibilities for pipeline operation analysis. More temporary measurement classes can be added as needed. How to access and display this type of real-time data will be introduced in volume 2 of this book series.

3.4.11 PSDM Implemented as Pipeline Spatial Database

PSDM defines a data framework that describes the pipeline. However, in order for PSDM to contain real pipeline data, Core Classes and Entity Classes of PSDM need to be instantiated and object attribute values of these classes should be filled. This is the implementation process of PSDM. As mentioned previously, PSDM aims to implement as a spatial database of Geodatabase and has referred to the object model design of Geodatabase at the design stage. There are several ways to implement PSDM as a Geodatabase, but the quickest and most effective method is to first model the PSDM object through UML (Unified Modeling Language) [47], and then generate the corresponding dataset of Geodatabase through ArcGIS Case Tools extension [48]. UML is a standard developed by OMG (Object Management Group) to express object-oriented analysis and design decisions. Case Tools, an independent extension of ArcGIS Desktop, has the following main functions:

(1) Providing related Geodatabase models for UML modeling directly in Microsoft Visio;
(2) Semantic checking of the modeling; and
(3) Reading the XMI (XML-based Metadata Interchange) file exported by UML and generating a Geodatabase. XMI uses a subset of the Standard Generalized Markup Language—Extensible Markup Language (XML), to provide programmers and other users with a standard way to exchange metadata information. XMI is also proposed by Object Management Group (OMG). XMI aims to help programmers using UML and different languages and development tools exchange data models with each other.

The steps of implementing PSDM as a spatial database of Geodatabase through UML and Case Tools are as follows:

1. Perform UML modeling of PSDM. Open the Geodatabase object models provided by Case Tools through Visio, and perform UML modeling with PSDM design schemas.

2. Export the built UML model as an XMI file. For Visio 2007, click “Macro”–“ESRIExportToXml”–“ESRIExportToXml”.
3. Perform a semantic check. For Visio 2007, click “Macro”–“ESRI”–“Semantics Checker”.
4. Generate a Geodatabase. Run ArcCatalog, click “Customize”—“Customize Mode”–“CASE Tools”, and add the “Schema Generation Wizard” command to the toolbar. Then, create a File Geodatabase or Personal Geodatabase, click the new Geodatabase, click the “Schema Generation Wizard” on the toolbar, import the XMI file generated in the previous step, set the projection and other information, and generate an empty Geodatabase.
5. Fill in or import information about specific pipeline elements.

In the above implementation steps, there are several points to note.

(1) Set the EventID field value for the elements of PSDM. Each class of PSDM has a field EventID that uniquely identifies a PSDM element. The data type of EventID is GUID (Globally Unique Identifier), which is actually 16-bit integer data, and theoretically unique for each GUID. The steps of generating EventID field value are as follows:

    1. Right-click on a layer under Table Of Contents on the left side of ArcMap and click Open Attribute Table;
    2. Right-click the EventID column in the opened layer attribute sheet and select Field Calculator to open the Field Calculator window;
    3. Check Python in the group of Parser, check Show Codeblock, and enter the following code under Pre-Logic Script Code;

```
def ID():
    import uuid
    return '{' + str(uuid.uuid4()) + '}'
```

    4. Enter "ID()" under "EventID=" and click OK to complete the generation of EventID.

Since the existing Centerline often does not have linear referencing measurement, if the pipeline data is imported from the existing data it is necessary to first set linear referencing measurements for these existing Centerlines before importing into PSDM. Steps are as follows:

1. Open the Centerline attribute sheet and add two fields, BeginMeasure and EndMeasure (the data type is Double). These two fields will hold the starting point linear reference measurement and the ending point linear reference measurement for each Centerline.
2. Set the value of BeginMeasure at 0 by the Field Calculator.

3. Right-click the EndMeasure column in the attribute sheet, select Calculate Geometry, select the Length in the list of Property, select Use coordinate system of the data source, select a unit from the list of Units, and click OK to complete the setting.
4. Activate ArcToolbox, click Linear Referencing Tools—Create Routes, open Create Routes, set Input Line Features as the layer completed in the previous step, set Route Identifier Field, set Output Route Feature Class, and set Measure Source to TWO_FIELDS from-Measure to BeginMeasure, and To-Measure to EndMeasure. Click OK to complete the setup. At this point, the newly generated layer has the linear referencing measurement.
5. Add the necessary fields to the newly generated layer according to the attributes of StationSeries, click ArcToolbox—Data Management Tools—General—Append, and import the newly generated layer into the StationSeries layer. At this point, the import from existing data to the PSDM StationSeries class has been completed.

In addition to importing the Centerline to PSDM StationSeries, if the pipeline data is imported from the existing data, it is also necessary to import the vertices of the Centerline to PSDM ControlPoint. Steps are as follows:

1. Click ArcToolbox—Data Management Tools—Features—Feature Vertices to Points, and export vertex data from the current Centerline;
2. Add the necessary fields for the vertex data according to the attributes of the StationSeries class, and set the linear referencing measurements of each vertex; and
3. Use the Append tool (ArcToolbox—Data Management Tools—General—Append) to import vertex data into the ControlPoint layer of PSDM.

If it is necessary to import the remaining pipeline online features, first use the Locate Features Along Routes tool (ArcToolbox—Linear Referencing Tools—Locate Features Along Routes) to set linear referencing measurements for online features, and then import the corresponding layer to PSDM through Append tool (for online feature stored in the event table) or Make Route Event Layer tool (ArcToolbox—Linear Referencing Tools—Make Route Event Layer) (for online feature stored as geometry features).

In PSDM, a feature with geometric fields can be stored in a feature class, in which case geometric attributes of the feature appear in the form of geographic reference coordinates. Features can also be stored in the event table, in which case geometric attributes of the feature are dynamically generated based on StationSeries and measure values. Both implementation methods have advantages and disadvantages, respectively. The implementation of feature class has the advantage of good display performance. Features can be quickly displayed on GIS software and directly edited by it. A disadvantage of using this method is that the feature position cannot be automatically updated when the corresponding StationSeries and measure values change.

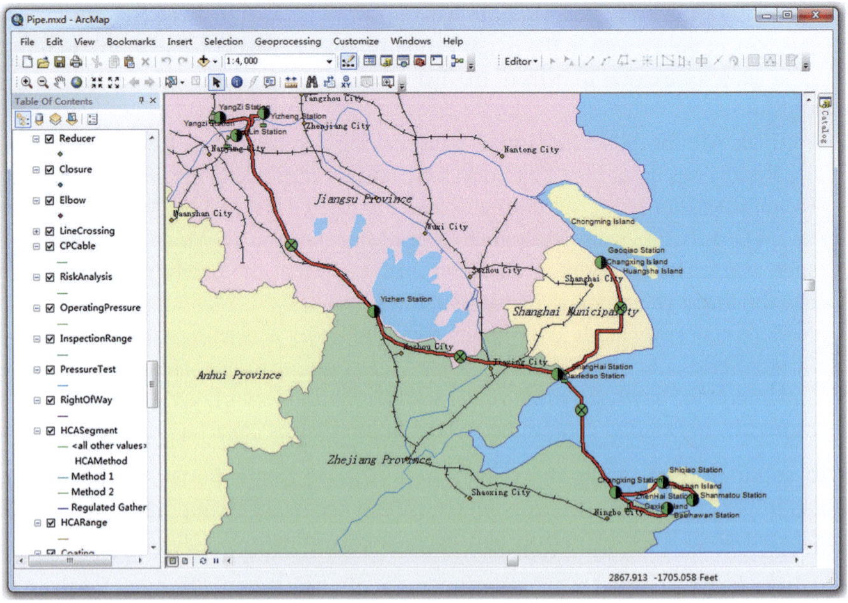

Fig. 3.17 PSDM implementation of Yong-Hu-Ning oil pipeline

Advantages and disadvantages of implementing the event table are the opposite. Since geometric attributes of the feature are dynamically generated based on StationSeries and measure value, with the implementation by event table, the feature position can be automatically updated when the corresponding StationSeries and measure value change.

Therefore, in the process of implementing PSDM as Geodatabase, it is necessary to consider the balance between the implementation of PSDM class as a feature or an event table. In general, geographic coordinates of pipeline entities do not change or do not change frequently, such as oil station. Instead, they can be implemented as features. Pipeline entities or events whose geographic coordinates often change, such as leakage, can be implemented as the event table.

By following the steps listed above, the pipeline data can be stored in Geodatabase spatial database with the logic and structure defined by PSDM, thus completing the construction of the pipeline spatial database. The author's implementation of PSDM using Yong-Hu-Ning oil pipeline data is shown in Fig. 3.17.

References

1. Yuan B, Shao J (2006) Geographic information system foundation and practice. National Defense Industry Press
2. Wu L, Liu Y, Zhang J, Ma X, Wei Z, Tian Y (2017) GIS: principles, methods and applications. Science Press
3. Zhang S (2001) Analysis of data model in GIS. Comput Eng Appl 37(8):122–124
4. Zeiler M (2010) Modeling our world: the ESRI guide to geodatabase design. 2 edn. Esri Press
5. Chen J, Zhang S (2003) The third generation geography data model and its realization. Geomat Spat Inf Technol 26(1):9–12
6. Liu Y, Jiang Y (2009) The study of geodatabase based on object-oriented technology and standard relational database. Geomat Spat Inf Technol 32(4):139–142
7. Guo X, Zhang H (2008) The third geographical data model—geodatabase and its implement. J Taiyuan Univ Sci Technol 29(1):5–8
8. Dong K, Fang Y (2004) Concept of spatial database model and its architecture research. Geomat World 2(2):8–16
9. Song Y, Wan Y (2004) The new spatial data model of Arc/Info 8 geodatabase. Bull Surv Mapp 11:31–33
10. ESRI. What is a geodatabase? https://desktop.arcgis.com/en/arcmap/latest/manage-data/geodatabases/what-is-a-geodatabase.htm. Accessed 06 May 2018
11. Yang X (2008) Feature analysis of geodatabase data model. Geomat Technol Equip 10(3):32–34
12. ESRI. Geodatabase. http://desktop.arcgis.com/en/arcobjects/latest/java/3400ece2–6f95-4c14-8d21-39aa6e6cd75b.htm. Accessed 06 Aug 2018
13. ESRI. Understanding ArcSDE. http://downloads.esri.com/support/documentation/sde_/706understanding_arcsde.pdf. Accessed 06 May 2018
14. Liu R, Liu N (2006) ArcGIS development collection: from entry to proficiency. Science Press
15. ESRI. What is ArcSDE? http://edndoc.esri.com/arcobjects/9.2/NET_Server_Doc/manager/geodatabase/administering_a-557706548/what_is_arcsde_qst_.htm. Accessed 06 May 2018
16. Qing T, Gong J, Li Q, Shen G, Yang M, Liu Y, Li J, Cao Q (2015) The application of linear reference in selected for the oil and gas pipeline. Geomat Spat Inf Technol 38(06):198–199+202
17. Scarponcini P (2002) Generalized model for linear referencing in transportation. GeoInformatica 6(1):35–55
18. Xiao J, Gao G, Peng T, Yin C, Lin T (2010) Design and realization of public transport inquiry system based on dynamic segmentation technology. Urban Geotech Investig Surv 02:58–61
19. Zhao L, Xie S, Tang J (2005) An implementation of dynamic segmentation in MapInfo. Acta Geod Cartogr Sin 34(2):175–178
20. Wang G, He Z, Zhang X, Xu H (2008) The research of the linear referencing system and dynamic segmentation in the highway GIS. Sci Surv Mapp 33(3):181–183
21. Gui Z, Yan L, Yan M (2003) The application of linear reference system and dynamic segmentation in GIS-T. Comput Eng Appl 39(09):208–209+215
22. Qiao Y, Wu H (1995) Studies on dynamic segmentation in GIS. Environ Remote Sens 10(03):211–216
23. Tong X, Yang D (2001) Development of a new linear referencing system data model. J Tongji Univ (Nat Sci) 29(4):410–415
24. Cadkin J, Brennan P (2002) Dynamic segmentation in ArcGIS. http://www.esri.com/news/arcuser/0702/files/dynseg.pdf. Accessed 11 Jan 2018
25. Gao Y, Liu Y, Wu L (2003) Dynamic segmentation and its implementation in GIS network analysis. Geogr Geo-Inf Sci 19(4):41–44
26. Guo L, Wang Y (1999) Dynamic segmentation technology in transportation geographic information system (GIS-T). Shanghai Highw 4:36–38
27. Supermap. Overview of dynamic segmentation. http://support.supermap.com.cn/datawarehouse/webdochelp/idesktop/features/dynamicseg/aboutdynamics.htm. Accessed 22 Feb 2018

28. Li Z, Wang J, Brook R, Kamelger A, Easton R (2016) Pipeline data model promoting data requirement for the oil & gas pipeline integrity management. Oil Gas-Eur Mag 42(2):86–90
29. Jin J, Wang X, Ding C (2018) Data model for production automation of long-distance oil and gas pipelines. Oil Gas Storage Transp 37(04):454–461
30. Cheng Z, Li C, Guo C (2006) China pipeline data model. China Investig Des 10:46–48
31. Jia Q, Wang Q, Wan Q, Bai J (2008) Data model for the realization of long-distance pipeline integrity management and its application. Geo-Inf Sci 10(5):593–598
32. Zhou L, Jia S (2014) Progress and development in the informatization of pipeline integrity management. Oil Gas Storage Transp 33(6):571–576
33. Bever KD (2003) Analysis of pipeline data standards. https://www.prci.org/Research/55984/18503.aspx. Accessed 06 Jan 2018
34. PODS (2013) PODS 6.0 Model diagram. https://www.pods.org/wp-content/uploads/2015/06/PODS-6.0-Logical-Models1.pdf. Accessed 20 May 2018
35. PODS. PODS model. https://www.pods.org/pods-model/what-is-the-pods-pipeline-data-model/. Accessed 16 May 2018
36. PODS. What is a PODS-compliant model? https://www.pods.org/about/faq/. Accessed 29 Feb 2018
37. Brush R (2002) The PODS data model. In: Proceedings of the 4th international pipeline conference, September 30, 2002–October 3, 2002, Calgary, Alta., Canada, 2002. Proceedings of the international pipeline conference, IPC. American Society of Mechanical Engineers, pp 1277–1282
38. Veenstra P (2007) Introduction to the ArcGIS pipeline data model (APDM). http://proceedings.esri.com/library/userconf/pug07/papers/breakouts/overview-of-apdm.pdf. Accessed 11 May 2018
39. Smith J (2003) ArcGIS pipeline data model (APDM). http://proceedings.esri.com/library/userconf/pug03/docs/apdm_workshop.pdf. Accessed 11 May 2018
40. APDM (2010) ArcGIS pipeline data model version 5.0.1—Reference Guide
41. Veenstra P (2008) Valid implementation of APDM. https://library.esri.com/docs/120328.pdf. Accessed 11 May 2018
42. Jia Q (2018) The application of ArcGIS in pipeline industry. http://www.docin.com/p-20351938.html. Accessed 06 Aug 2018
43. Li Y, Tan X, Zhou L, Yu H (2009) Applying APDM To pipeline integrity management at PetroChina. Pipeline Gas J 236(3)
44. Brook R, Wilson S, Veenstra P (2009) Pipeline data management. https://s3.amazonaws.com/webapps.esri.com/esri-proceedings/pug08/papers/tuesday/512_pipeline_data_management_by_rob_brook.pdf. Accessed 03 March 2018
45. PODS. PODS Spatial 6.0. https://www.pods.org/pods-model/pods-esri-spatial-geodatabase/. Accessed 06 May 2018
46. Gharabagh MJ, Asilian H, Mortasavi SB, Mogaddam AZ, Hajizadeh E, Khavanin A (2009) Comprehensive risk assessment and management of petrochemical feed and product transportation pipelines. J Loss Prev Process Ind 22(4):533–539
47. Group OM (2017) About the unified modeling language specification version 2.5.1. https://www.omg.org/spec/UML. Accessed 06 May 2018
48. ESRI. Using case tools and uml to create a geodatabase. schema. http://desktop.arcgis.com/zh-cn/arcmap/10.3/manage-data/geodatabases/using-case-tools-and-uml-to-create-a-geodatabase-schema.htm. Accessed 11 May 2018

Chapter 4
Component GIS, ArcObjects and ArcGIS Server

Abstract Geographic information systems can be divided into two types according to content, function, and role: the tool-type geographic information system and the applied geographic information system. The pipeline GIS system implemented in this book is an applied GIS. The main development methods of the applied geographic information system are introduced, then analyzed and compared. The compared conclusion is that the Component GIS-based method is suitable for the development of applied geographic information system. The author briefly summarizes the concept, roles, characteristics, and component models of Component GIS. The technical characteristics of Microsoft's Component Object Model (COM) are emphatically introduced. The most widely used GIS component library is ArcObjects. The process and method of applying ArcObjects to network environment through ArcGIS Server are elaborated in detail.

Keywords Component GIS · Component Object Model · ArcObjects · ArcGIS Server · Network environment · Programming interfaces

4.1 Research on Development Methods of Pipeline GIS Functions

The geographic information system can be divided into two types according to content, function, and role: the tool-type geographic information system and the applied geographic information system. The tool-type geographic information system, also called the geographic information system platform, is a platform with basic functions of GIS and can be used by other systems or customized by users. So far, there have been many commercialized tool-type geographic information systems, such as ArcInfo, MapInfo, MGE, and SuperMap. The applied geographic information system is a system designed to solve one or more types of practical application problems according to the user's needs or application objectives. In addition to basic GIS functions, it also has application models and methods for solving distribution patterns, distribution characteristics, and interdependence of geospatial features and spatial information. The applied geographic information system can be specially designed and developed for a professional department. Such a system is clearly targeted and

Z. Li, *Pipeline Spatial Data Modeling and Pipeline WebGIS*,
https://doi.org/10.1007/978-3-030-24240-4_4

highly specialized [1]. The Web-based Digital Pipeline system belongs to an applied geographic information system with oil and gas long-distance pipeline as the application object.

With the expansion of GIS application fields, the development of applied GIS becomes increasingly important. How to effectively develop a GIS that meets the needs and has a user-friendly interface for different application objectives is a concerned issue for GIS developers. At present, there are three main development methods of the applied geographic information system: independent development, customized development based on GIS platform, and customized development based on GIS components [2, 3].

(1) Independent development: All algorithms, from spatial data collection and editing to data processing analysis and result output. They are designed independently and implemented by developers selecting some programming tools and languages, such as VC and Delphi, and programming on a certain operating system platform without relying on any GIS tool software. The advantage of the method is reduced development costs without depending on any commercial GIS tools. For most developers, the limitations in capabilities, time, and financial resources make it difficult to functionally compare their products with commercial GIS tools.
(2) Customized development based on the GIS platform: The application system development is based on the GIS platform software, most of which provides a scripting language for user-customized development. For instance, ESRI's ArcInfo provides Python, while MapInfo's MapInfo Professional provides MapBasic. Users can use these scripting languages to develop their own application programs for different application objects by taking the original GIS software as a development platform. It is helpful in saving time and effort by the method, but the scripting language for customized development, as a programming language, is relatively weak. Meanwhile, the developed system cannot be separated from the GIS platform software and is executed with interpretation, which results in low efficiency.
(3) Customized development based on GIS components: The implementation of various functions of applied GIS by embedding GIS components in application programs through visual programming tools.

It is quite difficult for independent development. Due to the limitations of the programming languages provided by GIS platforms, customized development method based on the GIS platform is also not an ideal method. Therefore, the customized development method based on Component GIS combined with GIS components and today's visual programming languages has become the mainstream of GIS application development. Its advantage is that it can make full use of GIS components' function of managing and analyzing spatial data, as well as advantages of visual programming languages' high efficiency and convenience. With the above strengths, not only can the development efficiency of the application system be greatly improved but the application developed by the visual software development tool also has better user interfaces, more powerful database function, and reliability, ease to migrate, and convenient to maintain. These advantages can be particularly obvious if the integrated development is carried out by using GIS functional components with OCX technology. Due to these advantages, component-based GIS development has become the mainstream of the current GIS application development.

4.2 Component GIS and Component Models

4.2.1 Concepts and Main Ideas of Component GIS

Component software technology has become one of the trends in today's software technology. In order to adapt to the technology trend, GIS has undergone revolutionary changes like other software, i.e., the transition from all the functions or software with customized development functions provided by software providers in the past, to components provided by software providers but customized by users themselves. Undoubtedly, Component GIS will have a huge impact on the entire GIS architectures and application patterns.

Component GIS is a GIS system (including fundamental platform and application system) that uses object-oriented and component-based technologies. Component GIS consists of a set of components that allow cross-language applications with certain standard communication interfaces. Communication between GIS components and between GIS components and other components, accomplished through standard communication interfaces, can be performed across programs and computers.

The basic idea of Component GIS is to divide the major functional modules of GIS into several functional components. Each component performs different functions. These components can be developed by different manufacturers with any language that supports component technology, and there is no special limitation in development environments. Between GIS components, and GIS components and other non-GIS components, it is easy to integrate them through visual software development tools to form the final GIS application. GIS components are like a stack of various building blocks which implement different functions (including GIS and non-GIS functions). An application system will be formed as needed by those "building blocks" that implement various functions [4, 5].

Based on the component object platform, Component GIS has standard interfaces and allows cross-language applications, thus making GIS software more configurable, extensible and open, more flexible to use, and more convenient for customized development. Component GIS can not only solve difficulties faced by traditional GIS in software development, application system integration and user learning but can also help to reduce costs and gain extensibility. Therefore, most GIS software companies in the world have begun to develop Component GIS as an important development strategy. Component GIS has been an important trend in the development of GIS today.

4.2.2 Characteristics of Component GIS

Appropriate abstraction of the functionality of GIS, in the form of components for developers to use, will bring many advantages unmatched by traditional GIS tools [5–7].

(1) Efficient and seamless system integration: The establishment of a system often requires integration of GIS data, basic spatial processing functions, and various application models. The system integration schema, largely determining the applicability and efficiency of the system, is often different for different application fields and different application developers. To sum up, there are four main modes of integration based on traditional GIS software. ① Between the GIS software and the application analysis model, a data exchange channel is established through file access. In this mode, the GIS software and the application analysis model are independent of each other, and the system integration is poor, which is not suitable for large and frequent exchange of data. ② The application analysis model is compiled directly by the customized development language provided by the GIS software. In this mode, it is hard to develop complex application models as most customized development languages provided by GIS cannot be compared with professional programming languages such as C++, VB, and Delphi. ③ Application models are developed by using professional programming languages and directly accessing the internal data structures of GIS. In this mode, it increases the difficulty of application development of directly accessing the GIS software data structures. ④ Rapid communication between GIS and application models is established through Dynamic Data Exchange (DDE). In this mode, GIS and application models are still separated. In a word, traditional GIS software has limitations in system integration by adopting any of the above system integration modes. Component GIS provides the ideal solution to the above problems. Component GIS does not depend on a certain development language. It can be embedded in a common development environment (e.g., Visual Basic or Delphi) to implement GIS functions. Professional models can be implemented by using these common development environments, as well as plugging in other professional model analysis controls. Therefore, using Component GIS enables efficient and seamless system integration.

(2) Flexible structure and low development cost: The software itself is becoming increasingly large, the interaction of different systems is poor, and the development of the system is difficult, due to the closed nature of the traditional GIS structure. Under the component model, each component centrally implements system functions that are most closely related to it. The Component GIS platform provides functions such as collecting, storing, managing, analyzing, and simulating spatial data. As for other non-GIS functions (such as relational database management and statistical chart production), special components provided by professional vendors can be used to help reduce GIS development costs. On the

other hand, Component GIS itself can be divided into multiple controls to perform different functions. Users can select the required components according to their actual needs to minimize costs as much as possible.

(3) No need for specialized GIS development language: Traditional GIS often has an independent customized development language that may lead to a learning burden for users and application developers. Moreover, with the customized development language provided by the system, development is often limited making it difficult to deal with complex applications. Component GIS, built on strict standards, only requires the implementation of GIS basic functions instead of additional GIS customized development language and developing interfaces according to component model standards. It helps to reduce the burden on GIS software developers and enhances GIS extensibility. For GIS application developers, they do not need to master the additional GIS development language, but be familiar with the general integrated development environment, as well as attributes, methods, and events of GIS controls. Therefore, they can complete development and integration of the application system. At present, there are many development environments to choose from such as Visual C#.net, Visual C++, Visual Basic, Delphi, and C++ Builder. They can directly become excellent development tools for GIS, and their respective advantages can be fully utilized. It is a qualitative leap compared to the traditional GIS specialized development environment.

(4) High performance: The latest GIS components are based on the 32-bit system platform and in the form of InProc direct call, so both the ability to manage big data and the processing speed are not inferior to traditional GIS software.

(5) Reusability: It is the most basic characteristic of component software and the initial driver of combining component technology and GIS technology. Compared with traditional reuse technologies (code segment reuse, class reuse, etc.), component reuse focuses more on the wide range of software reuse and ease of software reuse. GIS software components reuse should also focus on the reuse in specialized application areas combined with other non-computer fields.

4.2.3 The Most Commonly Used Component Model for Component GIS—COM

Component technology is a binary-based code reuse technology. In component technology, a component program that provides a service is generally called server, and a component program that requires service is called client. The core of component technology is the two-way communication between client and server. At present, there are three representative component models widely used in the industry, namely Common Object Request Broker Architecture (CORBA) [8] proposed by Object Management Group, and EJB (Enterprise JavaBean) [9] constructed by Sun based on J2EE, Component Object Model (COM) owned by Microsoft [1, 10]. Although there are several component models mentioned above, from the historical experience of the software industry, COM is the most widely used component model. The research application object of this book is Component GIS based on COM/ActiveX specifications. The COM system mainly includes three aspects: COM, DCOM, and ActiveX [1, 2, 7, 11].

COM is not an object-oriented language, but a source-independent binary standard that defines the specification of communication between components. It is the foundation of Object Linking & Embedding (OLE) and ActiveX. Its function is to enable various software components and applications to interact in a unified standard way. COM provides specifications of the interaction between components, as well as an interactive environment. What COM creates is a link between a software module and another software module. When the link is established, modules can communicate through a mechanism called "interface". COM is still essentially a client/server model. Client (usually the application) requests to create a COM object and manipulate the COM object through the interface of the COM object. Server creates and manages COM objects based on the client's request. The roles of client and server are not absolute.

COM objects include attributes (also known as states) and methods (also known as operations). The object state reflects its existence and is also a feature distinguishing it from other objects. On the other hand, the method provided by the COM object is the interface provided by the object to the outside world, and the client must obtain service through the interface. For COM objects, the interface is the only way to interact with the outside world, so encapsulation is a fundamental feature of COM objects.

The location of the COM object is transparent to client because it does not directly access the COM object. Client program creates and initializes the object through a global identifier. In the implementation of COM, each COM object is identified by a 128-bit Globally Unique Identifier (GUID), called Class Identifier (CLSID). The object identified by CLSID can be theoretically guaranteed global uniqueness.

Clients and COM objects interact with each other through the interfaces, so the definition of the interfaces between components is crucial. One of the core contents of COM specification is about the definition of the interfaces. Technically, an interface is a data structure that contains a set of functions through which the client code can call the functions of a component object. The interface defines a set of member functions, which is all the information exposed by component objects. The client program uses these functions to obtain services of component objects. The interface is also identified by a 128-bit global identifier (IID, Interface Identifier). The client obtains an interface pointer through IID, and then through the interface pointer client can call its corresponding member function.

COM standard consists of two parts, the specification and the implementation. The specification defines the mechanism for communication between components and does not depend on any particular language or operating system. Any programming language can use this component as long as it complies with the specification. The implementation part of COM standard is COM library, which provides some core services for the concrete implementation of the COM specification.

COM based on distributed environment is called Distribute COM, distributed component object model (DCOM). DCOM is a natural continuation of COM in distributed computing, providing an interoperable infrastructure for COM components distributed in different nodes of a network. It is the foundation of ActiveX.

ActiveX is a set of COM-based technologies that enable software components to interoperate in a network environment regardless of the language in which the component was developed. It is the interface standard for Microsoft's distributed computing platform. As a technology developed for Internet applications, ActiveX is widely used in all aspects of Web servers and clients. At the same time, ActiveX technology is also used to conveniently create common desktop applications.

As an important part of ActiveX technology, ActiveX control is a programmable, reusable COM-based object. Similar to the ActiveX objects, ActiveX controls interact with containers and other controls through attributes, methods, events, etc. The difference is that ActiveX controls usually have their own display interface. Some software development companies specialize in producing ActiveX controls for a variety of purposes, such as database access, data monitoring, data display, graphic display, image processing, and even 3D animation. GIS controls based on this type of ActiveX include MapX from Mapinfo, MO from ESRI, and SuperMap controls from SuperMap. These controls integrate most of the commonly used objects, attributes, and methods of GIS. With the help of common object-oriented visualization tools, it is easy to develop custom GIS products. It is one of the important directions of GIS development based on ActiveX component technology.

4.3 COM-Based GIS Component Library—ArcObjects

4.3.1 ArcObjects Overview

At present, most GIS software companies in the world have taken developing component software as an important development strategy. Intergraph Corporation claims to have entered the era of Component GIS, and its GeoMedia Component GIS software is part of its massive Jupiter program. MapInfo Corporation launched MapX. After the launch of MapObjects, ESRI had been dedicated to building a full Component GIS platform, ArcObjects [12, 13]. ArcObjects is by far the most complete and complex GIS component library. It is a set of COM components built on Microsoft's COM technology. Its version 9.1 has 3051 components and 3986 interfaces. ArcObjects provides functions ranging from basic map operations to advanced spatial analysis. Meanwhile, ArcObjects is the core of GIS software ArcGIS, launched by ESRI. It is the development platform for ArcGIS family members such as ArcMap, ArcGIS Engine, and ArcGIS Server. ArcObjects, included in ArcGIS products, is not an independent commercial software. Components in ArcObjects provide users with the ability to perform customized development and functional extensions. Developers can use the ArcObjects framework to improve ArcGIS performance or extend its applications, or to develop independent GIS programs. ArcObjects has powerful, advanced, and open GIS components, rich and flexible spatial features, and advanced and reasonable data structures that support a wide range of geographic data sources [2, 11]. Based on these advantages of ArcObjects, this paper uses ArcObjects as a GIS component library to develop the pipeline GIS system.

4.3.2 ArcObjects Functions

ArcObjects is a component object library for constructing the ArcGIS family of platforms. It covers most main functions of today's GIS. Since ArcObjects is developed strictly based on Microsoft's COM specifications, COM technology can be used to customize and extend its functions. The main features of ArcObjects include the following aspects [2]:

- Interactive display, query, and analysis of geographic features;
- Creating and analyzing thematic map based on attribute information;
- Spatial query and spatial analysis functions;
- High-quality map output;
- Basic image processing functions such as image correction, rotation, and reverse;
- Powerful editing functions including new creation, movement, modification, copying and pasting of spatial features, feature buffers, merging of features in the same layer, combination of features in different layers, intersection of features to generate new features, feature duplication, polygon segmentation and line segmentation to generate new features, feature deleting, line breaking, extensions, and annotation generating;
- Object editing and cancellation/repetition for short transactions in a single-user environment;
- Superposition of vector data and raster data; and
- Support for editing and analysis of network elements associated with logical networks.

4.3.3 ArcObjects Component Libraries

ArcObjects contains a large number of COM objects. Many functions of these COM objects are similar. In order to better manage these COM objects, ESRI puts these components into different component libraries. The core component libraries are shown in Table 4.1 [11, 12]:

Table 4.1 ArcObjects core component libraries

Component library name	Description
System	Provides some fundamental components that can be used by other libraries
SystemUI	Defines objects used by ArcGIS user interface components
Geometry	Contains the core geometric objects
Display	Contains component objects needed to display graphics on the output device
DisplayUI	Provides objects with a visual interface for auxiliary graphic display
Controls	Contains visual component objects that can be used in program development
Carto	Contains various component objects that serve the data display
GeoDatabase	Contains objects for operating the Geodatabase
DataSourcesFile	Contains objects for opening file format geographic data
DataSourcesGDB	Used to open geographic data whose data source is an Access database or any large relational database supported by ArcSDE
DataSourcesOleDB	Used to operate any database that supports OLE DB
DataSourcesRaster	Used to get raster data stored in multiple data sources
Output	Contains all the output objects of ArcObjects

4.4 Application of ArcObjects in Network Environment Through ArcGIS Server

In order to implement the network-oriented pipeline WebGIS system, one method is that various basic GIS functions can be developed from the beginning, such as zooming in and out, querying the map, but this method will lead to excessive development cost. Another method is to use Component GIS to develop various GIS services. It will not only reduce the development cost but also improve the development efficiency and make the developed application highly extensible.

Component GIS can also be used to develop web applications, provided that GIS components can run in a network environment. ArcGIS Server is the enterprise-level WebGIS development platform which enables ArcObjects components to run on the network and is used to develop ArcObjects-based web applications and Web Services.

4.4.1 ArcGIS Server and Its Programming Interfaces

4.4.1.1 Overview of ArcGIS Server

This book uses ArcObjects as the GIS component library to develop GIS functions of the pipeline WebGIS system, but the pipeline WebGIS is a Web-based GIS application, so it is necessary to solve application problems of ArcObjects in a network environment. ArcGIS Server of the ArcGIS family provides a solution for applying ArcObjects in a network environment.

ArcGIS Server is an enterprise-level GIS development platform based on ArcObjects. It is used to build a centralized, multiuser, and advanced GIS-enabled web application, Web Services, desktop GIS applications, and other enterprise-level GIS application. ArcGIS Server provides a wide range of Web-based GIS services, rich GIS functions, and industry-standard GIS applications, to support geographic data management, mapping, spatial analysis, data editing, and other GIS functions in a distributed environment [14].

GIS applications built on ArcGIS Server have characteristics such as low costs, extensibility, balanced Web Services, seamless integration with IT systems (DBMS, Web Server, Enterprise Application Server), and use of a standard network (LAN/WAN/Internet). The most prominent characteristic of the ArcGIS Server is the introduction of advanced GIS functions into the network environment. Prior to this, advanced GIS functions were only available from desktop software [15].

4.4.1.2 ArcGIS Server Architecture

ArcGIS Server is a distributed system with multiple components that can be configured on multiple computers. The various components of the ArcGIS Server system play a specific role in object management, load balancing [16], etc. ArcGIS Server can be summarized as a three-tiered structural model: GIS Server, Web server, and client [14, 15, 17]. The ArcGIS Server architecture is shown in Fig. 4.1.

(1) GIS Server: The host of Server Object. GIS Server publishes Server Object as a service for the client to use. It contains the core ArcObjects library and provides a flexible environment for ArcObjects to run on the server. Server Object is a software object that manages and provides GIS resource services such as maps or locators. Server Object itself is a coarse-grained ArcObjects component. It simplifies the programming models of a series of operations required to complete a task so that the client only needs to complete a collection of multiple tasks through calling one function, such as display of the map. Server Object exposes a high-level set of stateless methods and also provides a Simple Object Access Protocol (SOAP) interface to handle SOAP requests. Server Object can be served to users as Web Services. ArcGIS Server currently provides two Server Objects: one is the Map Server, which provides access to map documents; the other is the Geocode Server, which provides access to locators.

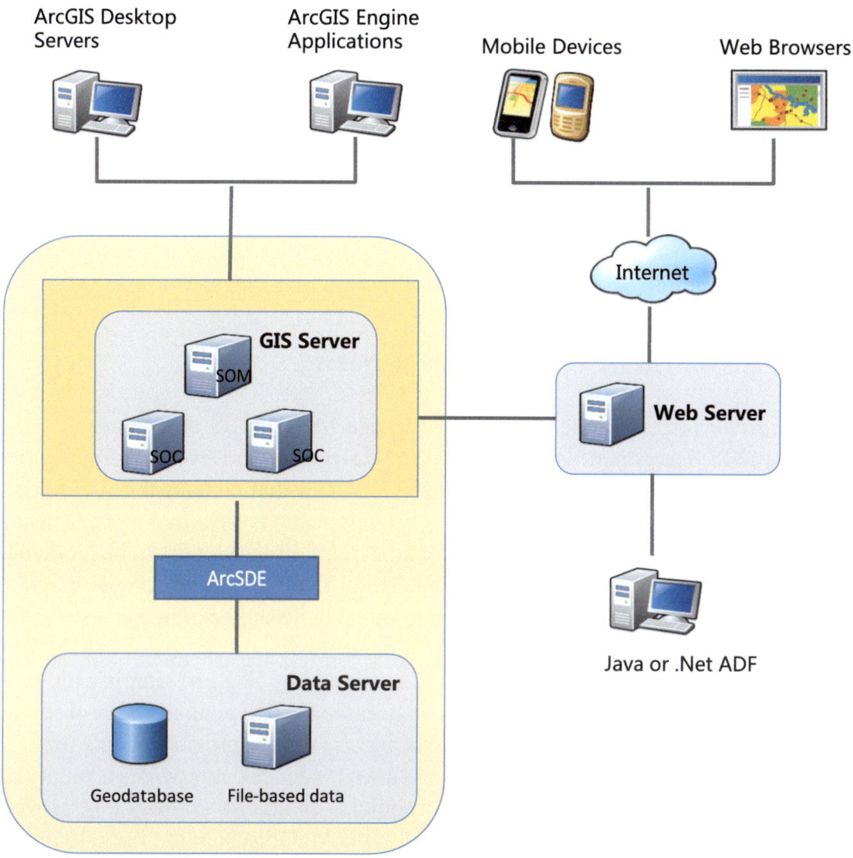

Fig. 4.1 ArcGIS Server architecture

The GIS Server includes a Server Object Manager (SOM) and one or more Server Object Containers (SOC).

SOM is a Windows/Unix service running on GIS Server. It manages Server Objects or Server Object groups distributed among one or more container servers, and is responsible for handing over external access to a process, balancing SOC loads, etc. It itself is an ArcObjects component and has permissions to use other ArcObjects components on the server.

SOM is connected to one or more SOCs, and Server Objects run on the SOC machine which means that SOC is the direct host of Server Objects. SOC is a process, which can accommodate one or more access instances of Server Object. When accessing a Server Object, the GIS Server will determine whether to establish an SOC according to the situation. All client requests are allocated by SOM and then completed by a certain SOC. According to different configurations, SOM and SOC can be deployed on different machines.

(2) Web server: Used to load web applications and Web Services. When it comes to GIS processing, these web applications and Web Services need to call ArcObjects objects running on the GIS Server.
(3) Client: includes web browsers, ArcGIS Desktop, and user-defined desktop applications based on ArcGIS Engine. The web browser connects to web application running on a Web server. The desktop system can connect to the GIS Web Service running on the Web server via HTTP or connects directly to a GIS Server via LAN/WAN.

4.4.1.3 ArcGIS Server Programming Interfaces

There are two programming interfaces available for WebGIS development based on the ArcGIS Server platform: Server API and Application Developer Framework (ADF). Server API is a collection of object libraries that contain the ArcObjects necessary to connect to a GIS Server and use applications such as Server Objects. Using Server API to build WebGIS applications requires developers to be proficient in the low-level ArcObjects object libraries. It is difficult to develop, but it can make full use of ArcObjects to develop a full-featured WebGIS application, and extend the GIS functions [18].

Web ADF is a development framework developed by ESRI to simplify the provision of GIS services such as map browsing on the Web. It contains a set of server controls and project templates built on ArcObjects, mainly providing tools for creating GIS applications. It supports .NET and Java languages. It provides a variety of visual controls and tasks, enabling users to quickly build GIS applications, as well as a number of class libraries that work well with ArcObjects in the background to perform complex GIS functions in a networked environment [19].

The method of developing with ADF can improve the efficiency of application development. However, it is not conducive to the customization and extension of GIS functions. Mixing the above two programming methods is a good choice, which can not only make full use of the powerful functions of ArcObjects to manipulate map objects but also avoid tedious issues for full use of the low-level ArcObjects components.

4.4.2 Implementing Network Application of ArcObjects Through ArcGIS Server

4.4.2.1 Mechanism for Remotely Calling ArcObjects Over Network

In the ArcGIS Server architecture, GIS Server hosts the ArcObjects components for managing and publishing map services and location services and is installed on the GIS Server. ADF, installed on a developer's machine, is a set of development components for developers. Both web applications and Web Services can use ADF. The GIS Server, Web server, and developer's machine can be the same machine or they can be installed separately.

Since the ArcObjects component set is on the server side, programming based on ArcGIS Server means manipulating the ArcObjects objects on the remote server. This is a big difference. Take ArcGIS Engine development as an example. The "new" keyword can be used to create an object on the local machine. This operation is always encapsulated in a process. The development based on ArcGIS Server means that a local object must call the remote ArcObjects object to implement a certain function. The local operation process and the remote operation process are actually two different processes, so it is necessary to solve the problem of how to communicate between the two processes.

ArcGIS Server uses Distributed Object Technology (DOT) to handle this problem. ADF provides a series of ArcObjects proxy objects. A proxy object is a local reference to a remote object. Its interface and methods are exactly the same as the remote object. In this way, the operation of the proxy object directly affects its proxy remote object.

When a web application is connected to the GIS Server, Server API used by the web application will call a proxy object to access the SOM object on the remote server and find the Server Objects managed by SOM through the SOM object. The specific process is as follows.

The web application sends user requests to SOM, and upon receiving the requests, SOM returns the web application one or more Server Object proxies running on the GIS Server. The web application uses the actual Server Object instances by using the Server Object proxies, just as the instance is in the process of the web application. In fact, the execution of Server Object instances occurs on the SOC of the GIS Server. The specific implementation of GIS functions is performed by Server Object instances in these SOCs. Therefore, the key to developing an ArcGIS Server-based web application is how to remotely call the functions of ArcObjects and Server APIs managed by the GIS Server.

4.4.2.2 Implementation of ArcObjects Remote Access in Network Environment

The core of ArcGIS Server is the ArcObjects component libraries, so the programming based on ArcGIS Server is substantially based on ArcObjects. The key to developing ArcGIS Server-based applications is how to remotely call ArcObjects in a GIS Server.

The general steps for ArcGIS Server development based on ArcObjects are as follows [15, 20]:

(1) Connect to GIS Server

The connection to GIS Server can be realized through the class GISServerConnection of ArcObjects. The GISServerConnection class implements the IGISServerConnection interface, and the public method "void Connect (string machineName)" of the

IGISServerConnection interface completes the connection to GIS Server. The parameter machineName is the host name of the GIS Server. The specific code is shown as follows:

```
IGISServerConnection connection=new GISServerConnection ();
Connection. Connect ("machine");
```

(2) Retrieve Server Object:

Server Objects are managed by SOM and run in Server Context, while a Server Context is a reserved space on the server running a set of Server Objects. The Server Context can be considered as a process managed by a server running Server Objects. Server Context provides a method, which exists as a running object, of creating ArcObjects in the same space and "process". Server Objects can be retrieved from the Server Context. The code is shown as follows:

```
IServerObjectManager som = conn.ServerObjectManager;
IServerContext context = som.CreateServerContext ("Pipeline", "MapServer");
IServerObject so = context.ServerObject;
IMapServer ms = so as IMapServer;
```

(3) Use Server Object:

A Server Object is a coarse-grained ArcObjects object. When you get a Server Object, other related ArcObjects objects can be accessed to implement specific GIS functions. The following code gets the first layer of a map service:

```
IMapServerObjects pMapServerObjs = ms as IMapServerObjects;
IMap map = pMapServerObjs.get_Map (ms.DefaultMapName);
ILayer layer =map. get_Layer (0);
```

(4) Release Server Context and Server Object:

After the use of Server Object, it should be released correctly in time to reduce resource waste. In the non-pooled case, once the context is released, Server Object obtained from the Server Context will no longer be used. The code is shown as follows:

```
context. ReleaseContext ();
```

References

1. Wu X (2015) Geographic information system design and implementation, 3 edn. Publishing House of Electronics Industry
2. Liu R, Liu N (2006) ArcGIS development collection: from entry to proficiency. Science Press
3. Jin J (2012) The principle and method of secondary development of GIS based on ArcGIS engine. Geomat Spat Inf Technol 35(3):46–49
4. Li Y-H, Deng H-Y, Zhao J-D, Chen Z-P (2005) Practice in comGIS development. Comput Eng Design (4):1090-1092
5. Kuang Z, Feng D, Yang Y (2009) Development of ComGIS technology. Geospatial Inf 7(4):114–117
6. Du J, Zhang Z, Liu Y (2004) Research and practice of ComGIS. J Geomat 29(4):18–20
7. Song G, Zhong E (1998) Research and development of components geographic information systems. J Image Graph 3(4):53–57
8. OMG (2015) CORBA BASICS. https://www.omg.org/gettingstarted/corbafaq.htm. Accessed 02 June 2018
9. Oracle. Enterprise JavaBeans Technology. https://www.oracle.com/technetwork/java/javaee/ejb/index.html. Accessed Feb 07 2018
10. Microsoft. The Component Object Model. https://docs.microsoft.com/en-us/windows/desktop/com/the-component-object-model. Accessed 02 July 2018
11. Jiang B (2006) ArcObjects development basics and techniques. Wuhan University Press
12. Zeiler M (2001) Exploring ArcObjects. ESRI
13. Lan X, Liu D, Wei R (2011) GIS application development based on ArcObjects and C#. Metallurgical Industry Press
14. ESRI (2004) ArcGIS server administrator and developer guide. ESRI Press
15. Wu G, Cong M (2006) Analysis of distributed gis application based on ArcGIS server. J Zhengzhou Inst Surv Mapp 23(1):52–55
16. Kang L, Fu J, Wang H, Cai J (2007) Development of WebGIS based on ArcGIS server. Water Resour Power 25(1):26–29
17. Lin G (2018) Introduction to ArcGIS Server. https://wenku.baidu.com/view/1aa73cd8be23482fb5da4c07.html. Accessed May Aug 2018
18. Zhao Z, Wang D, Zhou X (2007) The exploitation of WebGIS based on .Net, ArcObjects And ArcGIS server. Remote Sens Inf 1:76–80
19. ESRI. Developing Web applications using the Web ADF. http://help.arcgis.com/en/sdk/10.0/serveradf_net/conceptualhelp/index.html#/Developing_Web_applications_using_the_Web_ADF/000200000014000000/. Accessed 07 July 2018
20. Li Z, Li P, Ming W, Wang W (2010) Application of ArcGIS pipeline data model and GIS in digital oil and gas pipeline. In: International conference on geoinformatics, pp 1–5

Chapter 5
Pipeline WebGIS Implementation

Abstract In this chapter, the author explains the concepts, characteristics, and key technologies of Web Services, and summarizes the implementation methods and limitations of traditional WebGIS. To address the problems in the traditional WebGIS implementation methods, based on the analysis of the application mode of Web Services and Component GIS, this book proposes a WebGIS implementation method based on Web Services and Component GIS. Web Services that are used as the application framework to publish GIS functions are implemented by Component GIS, and then the GIS function published by Web Services, together with ArcGIS Server, is used to build the pipeline WebGIS system. This method can not only realize the GIS interoperability by using Web Services but also has the advantages of Component GIS, such as flexible structure, low development cost, high performance, and reusability.

Keywords WebGIS · Implementation · Component GIS · Web Services · ArcObjects · ArcGIS Server

After the pipeline spatial database is established, the remaining primary tasks to build a pipeline GIS include the following: input, update, query, analysis, and output of both pipeline spatial data and attribute data on the basis of GIS functions; implementation of simulation and analysis of pipeline operating conditions; and real-time pipeline data monitoring.

For the implementation of pipeline GIS, there are multiple development methods. An economical, powerful, and easy-to-implement development method is to be considered in this book, which mainly explores the Digital Pipeline system oriented to a network application. For this reason, WebGIS is adopted as the construction schema of pipeline GIS network application. At present, WebGIS can be implemented by various methods, but most of these implementation methods have common disadvantages of the inability to realize (1) the sharing of heterogeneous spatial data and GIS interoperability, (2) cross-platform, and (3) functional resources. Web Services, developed with XML as its core, provide new ideas for solving these problems. Therefore, this paper proposes a pipeline WebGIS construction schema based on Component GIS and Web Services.

Z. Li, *Pipeline Spatial Data Modeling and Pipeline WebGIS*,
https://doi.org/10.1007/978-3-030-24240-4_5

5.1 Research on WebGIS and Its Implementation Method

5.1.1 Overview of WebGIS

With the continuous development of Internet technology and the increasing demand for a geographic information system, it has become an inevitable development trend of geographic information system to publish spatial data through the Internet and provide users with spatial data browsing, query, and analysis functions.

WebGIS, namely Internet Geographic Information System, is a new type of geographic information system that has risen rapidly with the development of Internet technology. It uses the Internet for information publishing, data sharing, and communication and collaboration, and realizes functions of GIS such as online query and business processing. From the World Wide Web (WWW) nodes, Internet users can browse the spatial data on the WebGIS site through a browser to obtain data and functional services in the geographic information system, and perform various spatial retrieval and spatial analysis. WebGIS broadens the fields of geographic information resource application and provides a high degree of sharing GIS information. WebGIS has the functions of most traditional GIS software and the unique function of utilizing advantages of the Internet. Users can access remote GIS data and applications and perform GIS analysis on the Internet without having to install GIS software on their local computer [1–3].

At present, WebGIS has been widely used. Specifically, the application of WebGIS can be divided into spatial data publishing, spatial query retrieval, spatial model service, and organization of Web resources, etc.

5.1.2 Features and Benefits of WebGIS

Compared with the traditional geographic information system, WebGIS has its own special features which are mainly shown in the following aspects [4]:

- It is a network-based client/server system, while traditional GIS is mostly an independent stand-alone system.
- It uses the Internet to exchange information between the client and the server, which means that the transmission of information is global.
- It is a distributed system, where users and servers can be distributed in different locations and on different computer platforms.
- WebGIS, a cross-platform system, can access different platforms without having to care about what operating system (such as Windows, Unix, and Macintosh) the user is running. WebGIS has no restrictions on any computer or operating system. Users can access and use WebGIS as long as they have access to the Internet.

Web GIS has the following advantages [5]:

- broader access range,
- wide application,
- strong currency,
- independent platform,
- ease to use, and
- reduction in system cost on a large scale.

5.1.3 Implementation of Traditional WebGIS

There have been many ways to implement WebGIS so far. World Wide Web is based on the HTTP protocol, so any language, script, or technology that support it can be used directly for WebGIS application development. The commonly used WebGIS development methods can be based on two broad classes: the client-side mode and the server-side mode.

(1) WebGIS based on the server-side mode:

WebGIS based on server-side mode relies on the server-side GIS system to complete GIS analysis and result output. The client browser sends a service request to the Web server, and the server calls relevant GIS service programs after receiving the service request. The server accesses geographic data, executes the GIS function, and returns the execution result to the client browser in the form of a static web page or other forms. In this WebGIS mode, GIS data and GIS processing functions are located on the server side, and the client is only responsible for sending requests to the server and displaying the corresponding results sent back by the server.

The advantage of WebGIS development based on server technology is that the server can perform many complex GIS operations that are difficult to handle on the client side, and the system is easy to maintain and update. At the same time, the system has low requirements for the client. Any user who can browse the web page can obtain the GIS information. Even nonprofessional users can easily use various GIS functions.

The shortcoming of WebGIS development based on server technology is that all GIS-related operations are executed on the server side, meaning that every request from the client must be sent to the server for processing. This affects the performance and speed of the response causing the interaction between the system and the client to be poor.

Server-based WebGIS development technologies include Common Gateway Interface (CGI) and Server Application Programming Interface (Server API) [6–11].

(2) WebGIS based on the client-side mode:

WebGIS based on the client-side mode allows GIS analysis and GIS data processing to be performed on the client side. These GIS analysis tools and GIS data initially reside on the server. The user sends a request to the server for the GIS data and the GIS processing tools through a browser. The server transmits the required GIS data and the GIS processing tools to the client, the client receives them and then performs certain GIS data processing and analysis according to user's operation without the need of server participation. It has advantages of convenient operation, flexibility, and better performance as the required GIS data and GIS processing tools have been transmitted to the client.

The advantage of the client-side WebGIS development technology is that it enhances the client processing capability and the interactivity of the system, and reduces the amount of data processed by the server and the network transmission load.

The shortcoming of the client-side WebGIS development technology includes the high requirements for the performance of the client computer, difficulty in maintaining the system, and necessary training for users.

Currently, the client-side WebGIS development technology mainly includes Plug-in, ActiveX, and Java Applet [6–8, 12–14].

5.1.4 Limitations of Traditional WebGIS Implementation Methods

Since the birth of the first commercialized product of GIS in the 1980s, GIS has gradually formed an important computer application industry. Its development is closely related to the progress of computer technology. In particular, with the rapid advancement of the network technology represented by the Internet, GIS applications have been extended to the network environment. From the late 1990s, WebGIS systems have sprung up, which greatly promoted the wide application of spatial data. Most of these WebGIS systems use the implementation methods described above. However, due to some limitations of these implementation methods and some characteristics of GIS itself, the current WebGIS software technology still has some limitations in terms of universality, ease of use, extensibility, and interoperability, mainly in the following aspects [15–18].

(1) Cannot realize the sharing of heterogeneous spatial data and GIS interoperability:

GIS interoperability occurs in heterogeneous databases and distributed computing environments. Its basic feature is the mutual access to data and resources. GIS interoperability is the ability of GIS to access remote databases and processes in an integrative and transparent open environment. GIS interoperability requires that different applications (including software and hardware) dynamically call each other and that there are mutually accessible interfaces between different datasets. In the development of GIS, many relatively closed and independent systems have been formed. There is no uniform data format standard between these systems and the data storage and processing methods are also different. Even the seemingly identical operations have many differences due to the lack of a unified semantic description. When GIS is still at the application stage of project level or department level, the system interconnection and information sharing needs may not be particularly strong. Meanwhile, the limitation of the computer hardware processing capability and the interconnection and bandwidth of the network itself corresponding to this stage have also weakened and suppressed people's desires for data sharing and interoperability to a certain extent. When trying to build enterprise-level applications, or even establishing cross-industry, cross-administrative, and cross-border information system infrastructure, it is necessary to solve the problem of data sharing and GIS interoperability due to different formats of data, different GIS platforms, and the data distributed in different units and departments.

(2) Non-cross-platform:

Distributed application logic requires the use of distributed object models, such as Microsoft's DCOM, OMG's CORBA, and Sun's Remote Method Invocation (RMI). By using a distributed object model, developers can place services in remote systems. Nevertheless, these systems have a common shortcoming, that is, they all need a specific operating platform to provide basic network services and system services, and require tight coupling between the services provided by the client and the server. In other words, they require the use of the same type of basic structure. Such systems are often very fragile. If the execution mechanism at one end changes, the other end may no longer continue to run and can potentially crash. Therefore, the WebGIS platform built with these platforms cannot achieve cross-platform data access. It requires a more general model to abstract these distributed object models in order to achieve cross-platform on a higher level of abstraction.

(3) Cannot cross the firewall:

To ensure security, most enterprise platforms will choose to set up a firewall. Regardless of whether COM, DCOM, or CORBA technology is used, the communication code between the objects participating in the integration is in binary form, and communication can only be performed through a specific port such that it is very difficult for data to pass through the firewall.

(4) Functional resources cannot be shared:

For traditional GIS, the functional operations developed can only be used for dedicated applications and cannot be called by other systems. In this way, some common GIS operations have been repeatedly developed, which is a great waste of workloads and material resources.

In summary, the biggest shortcoming of the traditional WebGIS implementation method is the lack of sharing mechanism and GIS interoperability means, resulting in the inability to integrate spatial geographic information resources.

In the "Digital Earth" strategic conception, the sharing, interoperability, and integration of spatial geographic information are the key and foundation for its implementation. As a concrete application of the Digital Earth conception, Digital Pipeline should consider the interoperability of pipeline spatial data at the beginning of construction. The emergence of XML and Web Services technologies provide new ideas for solving the interoperability and integration of spatial data. In view of this, this book proposes a method to implement pipeline WebGIS system based on Web Services.

5.2 Research on the Implementation Methods of WebGIS Based on Web Services

For different WebGIS application systems, although the functional requirements of the final application systems are ever-changing; there are a lot of common parts in the development of these systems such as map operation functions and common spatial analysis functions. If the common parts of the map application system development can be abstracted and extracted, and published as the advanced GIS functions or services, it is possible to search for GIS functions and services like spatial data in the Internet environment.

The application of distributed computing and mobile computing has been increasingly popular with the development of network technology. People's demand for information services does not stay at the traditional concepts of active and passive services, but they hope that the system can provide all-round services at any time according to users' requirements. At the same time, the relatively tight coupling of WebGIS based on distributed component model technology such as CORBA and COM/DCOM has not been able to adapt to the looseness of Internet-based distributed computing requirements. It requires that WebGIS not only provides information publishing functions but also more importantly, provides information services. This kind of WebGIS will be a service-based WebGIS. WebGIS needs to be transformed from publishing functions to all-round services. Such network-based data or functional services are Web Services [19–21].

5.2.1 Web Services Concepts

At present, there is no unified definition of the Web Services concept. The following are the understandings and definitions of Web Services given by some organizations and well-known computer companies.

W3C: Web Services are a software application identified by URI whose interfaces and bindings can be defined, described, and discovered through XML components. Web Services support direct interaction with other software applications using XML-based messages via Internet-based protocols [22].

Microsoft: Web Services are Web components that are programmable and accessible through standard Web protocols [23].

IBM: Web Services, a kind of new web application branch, are self-contained, self-describing, and modularized applications that can be published, located, and called over the Web. Web Services can execute functions from a simple request to complex network processing. Once deployed, other web applications can discover and invoke the services [24].

Although there is no uniform definition of Web Services, the recognized Web Services technologies have the following in common:

(1) Web Services provide useful functions to Web users through standard Web protocols. The SOAP protocol is used in most cases.
(2) Web Services explain interfaces in detail, which enable users to create client applications and communicate with them. This description is usually included in an XML document called Web Services Description Language (WSDL).
(3) Web Services can be registered so that potential users can easily find them.

It can be said that Web Services are network resources that can be accessed through a URL address. Specifically, Web Services are applications built entirely on Internet standard protocols or specifications such as XML, and client applications can access them via standard protocols such as HTTP and SOAP. As can be seen from the above description, Web Services are, first and foremost, a kind of application logic that provides service. They are built on a widely accepted standard protocol so they can be supported by any system and development language. Finally, they are primarily used by program codes instead of the end users.

Web Services are mainly used for application collaboration between different system platforms so that various applications based on different platforms developed by different companies rely on them for connection and integration. Web Services will be the core of the next generation of distributed systems and have become the focus of today's IT community.

Many companies have adopted Web Services as their key technology for future development, such as Microsoft.Net and IBM Web Services Toolkit, which provide powerful support for Web Services. Among them, Microsoft.Net is the most representative and cutting-edge, providing the most extensive and comprehensive solution for the development and application of Web Services.

5.2.2 Web Services Features

The biggest advantage of Web Services is that they can adapt to multiple platforms, multiple systems, and multiple development languages on the network. Web Services' main goal is to build a common platform-independent and language-independent technology layer based on the existing heterogeneous platforms. Applications on different platforms rely on the technology layer to implement mutual connection and integration so as to provide users with a wide range of services. Users can choose the content, time, and method of obtaining information without having to browse through countless information islands to find the information needed as they did in the past. As can be seen from the concept of Web Services, they have the following features [18, 25]:

(1) Complete encapsulation: Web Services are objects deployed on the Web with good encapsulation. Users can only see the list of functions provided by a Web Services object.
(2) Loose coupling: This feature also derives from object/component technology. The caller keeps unknown when the implementation of a Web Service is changed. For the callers, as long as the interfaces of a Web Service are unchanged, any changes to its implementation are transparent to them even when the platform for its implementation is migrated from J2EE to .NET, or vice versa. The users do not need to know the implementation details of a Web Service.
(3) Using standard protocols: All public specifications of Web Services are fully described, transmitted, and exchanged with open standard protocols. These standard protocols have completely free specifications for implementation by any party. In general, most of the specifications will be published and maintained by W3C or OASIS.
(4) High integration capability: Web Services adopt simple and easy-to-understand standard Web protocols as component interface description specifications, which smooth out the differences between different software platforms. CORBA, DCOM, or EJB can interoperate through standard protocols to realize the high level of integration.
(5) Universal data exchange format: Any system that supports the open standard can understand Web Services by using an existing open standard rather than dedicated closed communication methods. Web Services use self-describing text-based messages to enable communication between autonomous and disparate systems. Web Services use XML to implement this function.
(6) Distribution: Web Services implementers and callers can be distributed throughout the network in the same process, in different machines, or in different machines in networks with completely different geographic locations. For an application system, multiple services required can also be distributed in different environments on the network.
(7) Cross-platform and cross-language: Web Services are built on a set of common protocols. The important ones are based on XML, which determines the difference between Web Services technology and traditional system integration technology. Web Services can run on various platforms, including typical Windows and Unix. Meanwhile, Web Services technology is beyond the boundaries of programming languages. Java, or C++, C#, and Visual Basic can be used to implement Web Services. Moreover, callers and implementers can use different programming languages.

5.2.3 *Web Services Architecture*

From a functional point of view, Web Services architecture is established based on the different operations of Web Services provider, Web Services requester, and Web Services registration agent [18]. The Web Services architecture model represented by roles is shown in Fig. 5.1.

As can be seen from the figure, the service-oriented Web Services architecture has three roles: service provider, service requester, and service registration agent. The service provider provides service functions to other services and users. Before providing the service functions, the service provider describes its services by Web Service Description Language (WSDL) and registers them with UDDI (Web Service Registration Specification) in the service registration agent. Service requester is the user of the services. It can use the service registration agent to find the required services. When the required services are found, the service registration agent provides the service description to it and binds it to the service provider. At this point, the service requester can send a request to the service provider to obtain the services. The service registration agent is capable of registering the published information of the service provider and the provided services and provides retrieval of the services. The three roles of service provider, service requester, and service registration agent are divided according to logical relationship. In actual application, roles may be crossed or interchanged. For example, a Web Service can be either a Web Services provider or a requester of another Web Service.

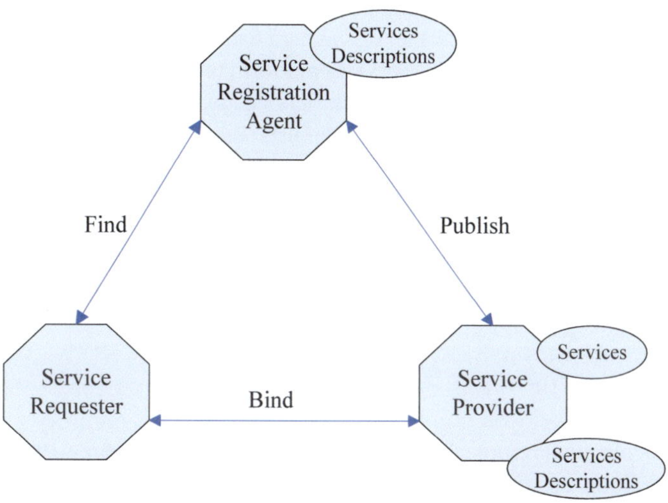

Fig. 5.1 Web Services role system model

5.2.4 *Key Technologies for Creating Web Services*

Key technologies for Creating Web Services include XML, SOAP, WSDL, and UDDI [26, 27].

5.2.4.1 XML

Extensible Markup Language (XML) is the core technology of Web Services standard and plays a vital role in implementing Web Services. It can be said that Web Services are built entirely on the basis of XML. XML is a standard way of describing data and exchanging data on a global scale. The biggest advantage of XML is that its data storage format is not restricted by the display format. It allows organizations and individuals to build a tag set that suits their needs. In addition, many complex data relationships perform well due to the self-describing nature of XML. XML provides a unified data format for Web Services. All messages and service descriptions use XML as the definition language to deliver messages and data flow so that data and messages from different platforms and environments can exchange.

XML Schema is a data modeling tool in the XML environment [28]. XML Schema is the recommended model by W3C and was officially released in May 2001. While using XML to describe data, it is also necessary to describe metadata of these data structures with a pattern language. The tool originally used for description is Document Type Definition (DTD). However, with the continuous application of XML, DTD can no longer meet the requirements. XML Schema is designed for targeting the shortcomings of DTD. It defines a set of standard data types and provides a language to extend the data types. The Web Services platform takes XML Schema Definition (XSD) as its data type system. Other Web Services technologies, including SOAP, WSDL, UDDI, and XML syntax, are also defined and described with XML Schema. XML Schema has become the standard communication tool in the XML world and is similar to UML's position in software design.

5.2.4.2 SOAP

Simple Object Access Protocol (SOAP) derives from the idea of XML-based RPC mechanism. In late 1999, with the joint efforts of Microsoft, DevelopMentor, and Don Box, it was developed into SOAP version 0.9. Its main purpose is to use the HTTP protocol to call remote COM objects across network and firewall restrictions. With the addition of companies such as IBM, SOAP is no longer limited to the Windows platform, and the protocol used is not just HTTP. SOAP has gradually evolved into a cross-platform, cross-language, and cross-protocol distributed object access technology. Currently, it has become a standard of W3C.

SOAP provides a simple, lightweight mechanism for exchanging structured and type information in a distributed environment in the form of XML. It is simple, does not require any object model, and can be done in any language. In fact, it defines a simple mechanism for representing application semantics by providing a package model with standard components and a mechanism for encoding data in a module, which enables SOAP to be used for various systems from messaging to RPC. SOAP is mainly composed of SOAP envelope, SOAP encoding rules, SOAP RPC representation, and SOAP binding.

The emergence of SOAP was inevitable. In the past, creating complex applications for application communication was done by using some object models, such as Microsoft's DCOM or CORBA. Nevertheless, these technologies have their own disadvantages when creating Web Services. If they are used to create distributed applications, it usually requires running the same distributed object model on both ends of the connection. However, the Internet does not guarantee that the other end of the connection is running a specific client and service software, and it is impractical for everyone to run COM objects or other objects. As an XML-based presentation layer protocol that does not rely on transfer protocols, SOAP facilitates the exchange of data between applications in the form of objects, smoothing the differences between component platforms. It is the solution to the aforementioned problem.

5.2.4.3 WSDL

A structured, understandable method is required to describe Web Services so that the description of Web Services can be automatically processed by programs. This is the basic guarantee for the instant assembly of Web Services. Web Services Description Language (WSDL) is exactly such a tool. It is a service description language similar to Interface Description Language (IDL) technology. WSDL satisfies this need by defining a set of XML-based grammars that describe Web Services as a collection of service access points that can exchange messages.

After implementing the network service function, the service provider describes the service with WSDL and publishes it on the network. After the user obtains the WSDL document of the Web Services, he can know the location of the Web Services, the methods they contain, the parameters of each method and the type of the return value, and use this information to execute the method call.

WSDL describes Web Services through messages exchanged between a service provider and a service requester. A message itself is abstractly defined and then bound to a specific network protocol and a message format. A message consists of a series of data items of the specified types, of which there are five main components:

- Types: defines the data types used in the WSDL definition, i.e., XML Schema types;
- Message: definitions of the input and output parameters of a set of messages;
- PortType: defines the operations of Web Services;

- Binding: describes the protocol, data format, security, and other attributes of a particular service interface; and
- Services: used to specify the URL and the provided interfaces of a particular service, including a set of port elements.

5.2.4.4 UDDI

Web Services are also resources on the Internet, so some methods must be used to find specific Web Services.

Universal Description Discovery and Integration (UDDI) specification defines a standard method for publishing and discovering information about Web Services. It is an XML-based specification for recording the business and services provided by websites. Drawing on the experience of XML and SOAP, UDDI, a layer defined on top of them, provides a mechanism for clients to dynamically publish and search for Web Services. Through the standard interfaces provided by UDDI, enterprises can publish their own Web Services for other enterprises to query and call. They can also query the description information of specific services and dynamically bind to the services.

The purpose of UDDI is to register and discover services, which serve users in the form of a UDDI online registry. UDDI's registry can be divided into two types: Public UDDI Registry and Private UDDI Registry. Public UDDI Registry provides global registration services. The registry does not refer to a registration site, but a generic term for all sites that provide registration services. Private UDDI Registry is established by a certain organization and is used by an independent enterprise within a certain scope. It does not provide global services.

5.2.5 *Web Services Usage Modes*

There are four general ways for users to use Web Services on the client:

(1) Directly call the Web Services. This method is used when the client develops an application. The user can directly call the service on the network in the program, such as calling the service that calculates the best path on the network when developing a GIS system.
(2) Local Web Services indirectly call remote Web Services—calls the Web Services from the called local services. It forms a nested structure. This method provides a solution for the interoperability of the client and services.
(3) Link to ASP.NET page for Web Services. This method is only applicable to Web Services developed with the .NET development environment. Users can type the address of a Web Service into a browser and see the interfaces of the service provided by .NET.

(4) User program downloads and runs the client executable programs for Web Services. This method is developed for the end user. Web Services themselves have been called by the programs, and the user can download the programs to use the functions provided by Web Services.

5.2.6 The Significance of Web Services for the Development of GIS

Web Services provide a platform-independent and language-independent mode of sharing data and services between machines in a network environment. Web Services working at the code level can be called by other software and exchange data with other software to eventually form an application system that can interact with the users. GIS system based on Web Services is expected at a higher level to solve the GIS problem of GIS data integration and sharing in a wide range, which cannot be sound solved by the GIS system based on Component GIS. Therefore, Web Services is an ideal technology for implementing GIS interoperability. In WebGIS based on the Web Services technology, each node is both a service provider and a service consumer, and both in the form of Web Services exposes the data and functions it can provide. At the same time, each node accesses data and functions provided by other nodes through SOAP. In this way, the dynamic analysis of large amounts of distributed spatial data and integration with other applications in a distributed environment can be realized. It can operate transparently between different systems and data and can be used for cross-platform applications and heterogeneous network interconnection. It has good human–computer interaction and data acquisition and interoperability to achieve maximum sharing of geographic information. It can be said that the GIS application based on Web Services is a major breakthrough in the implementation of GIS network integration and interoperability, and will become an important turning point in the development of GIS.

5.3 Implementation of Pipeline WebGIS Based on Web Services and Component GIS

The main purpose of Web Services for GIS is to abstract common functions of GIS, encapsulate them into Web Services, and publish them to the network for use by other applications so as to achieve interoperability of GIS. However, complex and non-common functions of GIS should not be encapsulated into Web Services because on one hand, it has little reusable value, and on the other, performance may also be affected due to their complexity. Therefore, the implementation of functions of GIS is divided into two cases. First, functions of GIS that need the transfer of a large amount of spatial data, or are implemented via complicated operations, or have no common features, such as roaming, query, input, and editing of pipeline spatial data, are implemented using ArcGIS Server plus ArcObjects. Second, functions of GIS with common features, such as editing of pipeline attribute data, and some common spatial analysis (such as overlay analysis) are encapsulated into Web Services and implemented using ArcObjects.

5.3.1 Serialization of ArcObjects Component Objects

The main reason for lacking interoperability of GIS is that organizations and manufacturers adopt their own GIS data format and software systems. One manufacturer's software cannot use another manufacturer's data format, and their software interfaces cannot communicate with each other. Therefore, it is necessary to unify the GIS data format and software interfaces or to make it easy to communicate between the two so as to achieve interoperability of GIS. Currently, OGC is developing relevant standards including the Geography Markup Language (GML). OGC intends to use GML as a standard for GIS data exchange, but GML has limited support for large amounts of spatial data. Judging from the current development, the standards developed by OGC are still not practical enough, and manufacturers will adopt their own GIS data formats and software systems for a long time. Considering ArcGIS's market share and influence in the world, Web Services are implemented using ArcObjects in this book. This causes limitations such as only allowing access to users of ArcObjects. However, in the design of Web Services, this book tries to adopt the general design principle. Standard geographic features (points, polylines, and polygons) and standard geographic reference coordinates are used as the return value and parameters of Web Services interfaces. In this way, even if the standard GIS data format and software interface appear in the future, it is only necessary to change the implementation within Web Services without changing the interface description of Web Services. Users who use these Web Services can still work normally, thus ensuring consistency of program access, which exactly reflects the perfect encapsulation of Web Services.

Since the return values and parameters of Web Services can only be the basic data type, ArcObjects is used to implement Web Services in this book. This means that ArcObjects data types will be used as parameters and return values of Web Services. Therefore, ArcObjects components and interface types should be serialized first (i.e., conversed to basic types), and then they can be used as return values or parameters of Web Services. Component objects that implement the IXMLSerialize interface can be serialized directly, and component objects that do not implement the IXMLSerialize interface but implement the IPersistStream interface can be indirectly serialized through the IXMLPersistedObject interface. Objects that do not implement the above two interfaces but can be in the form of relational tables, such as FeatureClass objects and feature objects, can be serialized after being converted to .Net DataTable objects. Normal fields of FeatureClass objects or feature objects can directly correspond to corresponding fields of DataTable objects. Geometric attribute fields of FeatureClass objects or feature objects can be serialized to common fields of DataTable objects by the first two methods because they implement the IXMLSerialize interface or the IPersistStream interface. In this book, an auxiliary class ESRIComSerializer written in C# language is used to implement the serialization of ArcObjects data types, which is defined as follows. ESRIComToString method is used to serialize ArcObjects objects into strings, ESRIComFromString method to construct ArcObjects objects from strings, ESRIFeatureClassToString method to serialize feature classes or features into strings, and ESRIFeatureClassFromString method to build feature classes or features from strings. The detailed implementation process of the latter two functions is not listed in this paper due to their long program codes.

```
public class ESRIComSerializer
{
    public static string ESRIComToString (object pESRIComObject)
    {
        string mSerialized = null;

        if (pESRIComObject is IXMLSerialize)
        {
            IXMLSerializer pXMLSerializer = new XMLSerializerClass () ;
            mSerialized=pXMLSerializer.SaveToString(pESRIComObject, null, null);
            Form1.mESRISerializeType = ESRIComSerializeType.XMLSerializeType;
        }
        else if (pESRIComObject is IPersistStream)
        {
            IXMLPersistedObject xmlWrapper = new XMLPersistedObjectClass () ;
            xmlWrapper.Object = pESRIComObject;
            IXMLSerializer xmlSerializer = new XMLSerializerClass () ;
            mSerialized = xmlSerializer.SaveToString (xmlWrapper, null, null) ;
            Form1.mESRISerializeType = ESRIComSerializeType.PersistStreamType;
        }
        return mSerialized;
    }
```

```
    public static object ESRIComFromString (string pSerializedString, int
pSerializeType)
    {
        object ESRIComObject = null;

        IXMLSerializer xmlSerializer = new XMLSerializerClass () ;
        if (pSerializeType == 1)
        {
            ESRIComObject = xmlSerializer.LoadFromString (pSerializedString, null,
null) ;
        }
        else if (pSerializeType == 2)
        {
            IXMLPersistedObject xmlWrapper = xmlSerializer.LoadFromString
 (pSerializedString, null, null) as IXMLPersistedObject;
            if (xmlWrapper != null)
            {
                ESRIComObject = xmlWrapper.Object;
            }
        }
        return ESRIComObject;
    }

    public static string ESRIFeatureClassToString (object pESRIComObject)
    {
    }

    public static object ESRIFeatureClassFromString (string pSerializedString)
    {
    }
}
```

5.3.2 *Implementation of Roaming, Query, and Editing Functions of Pipeline Spatial Data*

Before implementing network-based applications of pipeline spatial data, pipeline data should be published on the network. The pipeline spatial data model designed in this book is implemented as an SDE Geodatabase and stored in SQL Server Database. The data stored in the database should be published on the network as follows: ① register the pipeline-related layers as versioned in the database; ② open ArcMap to load the pipeline data; ③ save the loaded pipeline data as an mxd file; ④ publish this mxd file as a Map Server (i.e., Map Service) via ArcCatalog or ArcGIS Server Manager. After the pipeline data is published as a map service, the service can be connected through various channels (including web applications, ArcMap, and ArcGIS Engine programs), and various GIS functions can be implemented through ArcObjects. Please refer to the steps of connecting to the Map Server in 4.4.2.2.

As described above, the roaming, query, input, and editing functions of pipeline spatial data are implemented via ArcGIS Server plus ArcObjects rather than Web Services. The implementation of editing of spatial data is taken as an example as follows.

The input and editing of pipeline spatial data can be implemented through EditTask or the custom tool of ArcGIS Server ADF. EditTask provides a number of predefined tools that allow users to edit spatial data without writing code, but it is not flexible enough. The custom tool can control the editing process. The editing of spatial data is implemented through the custom tool in this book.

The custom tool of ArcGIS Server provides a mechanism to customize different behaviors on both the client and the server. Behaviors on the client include clicking, pulling a curve on the interface with the mouse, and pulling out a rectangle with the mouse. For these behaviors, the server can implement different GIS functions. For example, when the user clicks on the interface, the server can add a point feature to the database, or zoom in on the current layer. Users have great flexibility in the implementation of GIS functions through this customization mechanism.

Behaviors on the client can be configured on the corresponding custom tool controls of the toolBar. The corresponding behaviors on the server must be customized by implementing the IMapServerToolAction interface. The specific GIS functions can be implemented using the ServerAction method of the interface. This book mainly uses three custom tools to realize the input of pipeline spatial data, which are point tool, polyline tool, and polygon tool. The function of the three tools is to add point features, polyline features, and polygon features on the current layer, respectively. The implementation of the three custom tools on the server corresponds to three classes that implement the IMapServerToolAction interface, which are AddPointFeature, AddPolygonFeature, and AddPolylineFeature. Their implementation principles are basically the same—first converting the screen coordinates of the user action into geographic reference coordinates, and then creating corresponding ArcObjects features through the ServerContext context, and finally saving them to the corresponding layer. The detailed implementation of these three classes is as follows.

AddPointFeature class:

```
public class AddPointFeature :  IMapServerToolAction
{
    public AddPointFeature () {
    }
    #region IMapServerToolAction
    void IMapServerToolAction.ServerAction (ToolEventArgs args)
    {
        ESRI.ArcGIS.ADF.Web.UI.WebControls.Map mapCtrl;
        mapCtrl =  (ESRI.ArcGIS.ADF.Web.UI.WebControls.Map) args.Control;
        PointEventArgs pea =  (PointEventArgs) args;
        System.Drawing.Point screen_point = pea.ScreenPoint;

        MapFunctionality mapFunc =  (MapFunctionality)mapCtrl.GetFunctionality(0);
        MapResourceLocal mapResLocal = mapFunc.Resource as MapResourceLocal;
        IServerContext pSOC;
        IMapServer pMapServer;
        IMap pMap;

        pSOC = mapResLocal.ServerContextInfo.ServerContext;
        pMapServer = pSOC.ServerObject as IMapServer;
        IMapServerObjects pMapServerObjs = pMapServer as IMapServerObjects;
        pMap = pMapServerObjs.get_Map (pMapServer.DefaultMapName) ;

        ESRI.ArcGIS.ADF.ArcGISServer.MapDescription mapDesp = mapFunc.MapDescription;
        ESRI.ArcGIS.ADF.Web.Geometry.Point adf_map_point =
ESRI.ArcGIS.ADF.Web.Geometry.Point.ToMapPoint (screen_point,  mapCtrl.Extent,
mapFunc.DisplaySettings.ImageDescriptor.Width,
mapFunc.DisplaySettings.ImageDescriptor.Height) ;
        PointN ags_map_point =
ESRI.ArcGIS.ADF.Web.DataSources.ArcGISServer.Converter.FromAdfPoint(adf_map_point);

        ILayer pLayer = pMap.get_Layer (0) ;
        IFeatureLayer pFeatureLyr = pLayer as IFeatureLayer;
        IFeatureClass pFeatCls = pFeatureLyr.FeatureClass;
        IDataset pDataset = pFeatCls as IDataset;
        IWorkspace pWS = pDataset.Workspace;

        IWorkspaceEdit pWorkspaceEdit = pWS as IWorkspaceEdit;
```

```
            pWorkspaceEdit.StartEditing (false) ;
            pWorkspaceEdit.StartEditOperation () ;
            IFeatureBuffer pFeatureBuffer;
            IFeatureCursor pFeatureCuror;
            IFeature pFeature;
            IPoint pPoint;
            pFeatureBuffer = pFeatCls.CreateFeatureBuffer () ;
            pFeatureCuror = pFeatCls.Insert (true) ;
            pFeature = pFeatureBuffer as IFeature;
            pPoint = (IPoint) pSOC.CreateObject ("esriGeometry.Point") ;
            pPoint.X = ags_map_point.X;
            pPoint.Y = ags_map_point.Y;
            IGeometry pPointGeo = pPoint as IGeometry;
            pFeature.Shape = pPointGeo;
            pFeatureCuror.InsertFeature (pFeatureBuffer) ;
            pWorkspaceEdit.StopEditOperation () ;
            pWorkspaceEdit.StopEditing (true) ;
            mapCtrl.Refresh () ;
        }
        #endregion
    }
```

The detailed implementation of AddPolygonFeature and AddPolylineFeature is not shown here since their implementation methods are similar to that of AddPointFeature.

The complete interface of the pipeline WebGIS is shown in Fig. 5.2.

The query of pipeline data is also implemented by using ArcGIS Server plus ArcObjects because it involves the transfer of a large amount of data. The implementation mainly involves the IFeatureClass interface and the IQueryFilter interface, of which the core implementation codes are as follows:

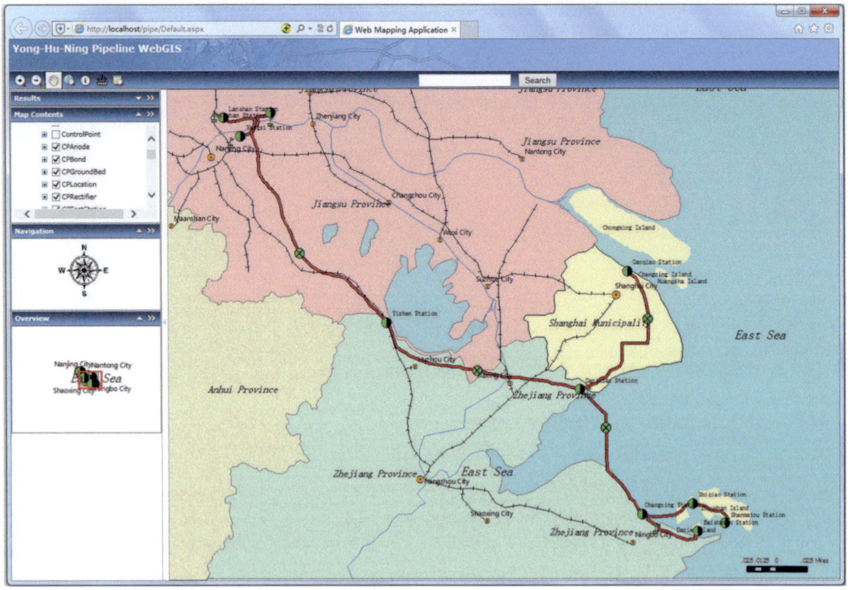

Fig. 5.2 Complete interface of the pipeline WebGIS

```
IQueryFilter qf= new QueryFilterClass () ;
IFeatureLayer featureLayer;
IFeatureClass featureClass;
IFeatureCursor featureCursor;

featureClass = featureLayer.FeatureClass;
qf.WhereClause = "Name=' Gaoqiao Oil Transmission Station'";
featureCursor = featureClass.Search (qf,  false) ;
while ( (feature = featureCursor.NextFeature () )  != null)
{
    //Process query results
}
```

Figure 5.3 shows the query results of the "Gaoqiao Station". The attribute data is displayed in the tree control on the left side of the interface, and the spatial features are displayed in the middle.

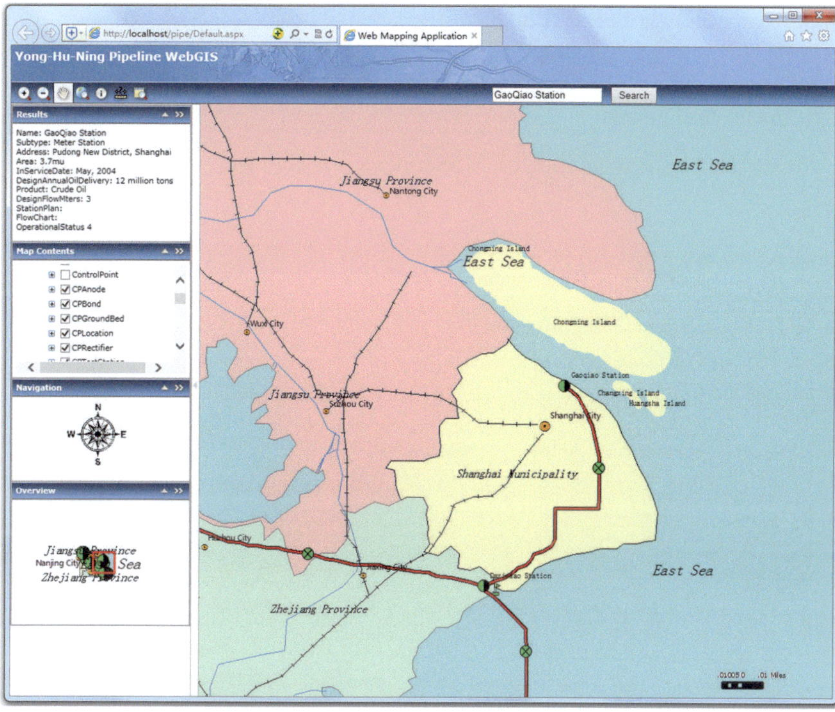

Fig. 5.3 Query of pipeline data

5.3.3 Implementation and Application of Commonly Used GIS Web Services for Pipelines

The pipeline GIS has some common functions, including attribute editing, overlay analysis, feature clipping, and buffer analysis. These common functions are implemented as Web Services in this book to achieve reuse and interoperability of GIS functions.

.Net C# is used to implement Web Services in this book. In C#, the class that contains Web Services must inherit from the System.Web.Services.WebService class. "[WebMethod]" should be added next to the Web Services method to indicate that the method is a Web Service. Parameters and return values of ArcObjects objects and interfaces should be firstly serialized and then passed. All Web Services in this book are implemented in a class GISServices that inherits from the System.Web.Services.WebService. The main methods of GISServices are as follows:

- SetFeatureValue—set the feature attribute value;
- Buffer—buffer analysis;
- Union—find the union of features;

- Intersect—find the intersection of features;
- Clip;
- Difference; and
- SymmetricDifference.

The detailed implementation of GISServices is as follows:

```
using System;
using System.Web;
using System.Web.Services;
using System.Web.Services.Protocols;
using ESRI.ArcGIS.Geodatabase;
using ESRI.ArcGIS.Geometry;

[WebService (Namespace = "http://tempuri.org/") ]
[WebServiceBinding (ConformsTo = WsiProfiles.BasicProfile1_1) ]
public class GISServices : System.Web.Services.WebService
{
    public GISServices () {
    }

    [WebMethod]
    public void SetFeatureValue (string strFeature, string strField , string value)
    {
        IFeature feature= ESRIComSerializer.ESRIFeatureClassFromString(strFeature);
        IFields fields = feature.Fields;
        int fieldIndex = fields.FindField (strField) ;
        feature.set_Value (fieldIndex, value) ;
    }


    [WebMethod]
    public string Buffer (string strFeature, double distance)
    {
        ITopologicalOperator pTopo;
        IGeometry pGeoBuffer;
        IGeometry pGeo = ESRIComSerializer.ESRIComFromString (strFeature) ;
        pTopo = (ITopologicalOperator) pGeo;
        pGeoBuffer = pTopo.Buffer (distance) ;
        return ESRIComSerializer.ESRIComToString (pGeoBuffer) ;
    }
```

```
[WebMethod]
public string Union (string strFeature1, string strFeature2)
{
    ITopologicalOperator pTopo;
    IGeometry pGeoBuffer;
    IGeometry pGeo1 = ESRIComSerializer.ESRIComFromString (strFeature1) ;
    IGeometry pGeo2 = ESRIComSerializer.ESRIComFromString (strFeature2) ;
    pTopo = (ITopologicalOperator) pGeo1;
    pGeoBuffer = pTopo.Union (pGeo2) ;
    return ESRIComSerializer.ESRIComToString (pGeoBuffer) ;
}

[WebMethod]
public string Intersect (string strFeature1, string strFeature2, int dimension)
{
    ITopologicalOperator pTopo;
    IGeometry pGeoBuffer;
    IGeometry pGeo1 = ESRIComSerializer.ESRIComFromString (strFeature1) ;
    IGeometry pGeo2 = ESRIComSerializer.ESRIComFromString (strFeature2) ;
    pTopo = (ITopologicalOperator) pGeo1;
    pGeoBuffer = pTopo.Intersect (pGeo2, dimension) ;
    return ESRIComSerializer.ESRIComToString (pGeoBuffer) ;
}

[WebMethod]
public string Clip (string strFeature, string strEnvelope)
{
    ITopologicalOperator pTopo;
    IGeometry pGeoBuffer;
    IGeometry pGeo1 = ESRIComSerializer.ESRIComFromString (strFeature) ;
    IGeometry pGeo2 = ESRIComSerializer.ESRIComFromString (strEnvelope) ;
    pTopo = (ITopologicalOperator) pGeo1;
    pGeoBuffer = pTopo.Clip (pGeo2) ;
    return ESRIComSerializer.ESRIComToString (pGeoBuffer) ;
}

[WebMethod]
public string SymmetricDifference (string strFeature1, string strFeature2)
{
    ITopologicalOperator pTopo;
    IGeometry pGeoBuffer;
    IGeometry pGeo1 = ESRIComSerializer.ESRIComFromString (strFeature1) ;
    IGeometry pGeo2 = ESRIComSerializer.ESRIComFromString (strFeature2) ;
    pTopo = (ITopologicalOperator) pGeo1;
    pGeoBuffer = pTopo.SymmetricDifference (pGeo2) ;
    return ESRIComSerializer.ESRIComToString (pGeoBuffer) ;
}
```

```
[WebMethod]
public string Difference (string strFeature1, string strFeature2)
{
    ITopologicalOperator pTopo;
    IGeometry pGeoBuffer;
    IGeometry pGeo1 = ESRIComSerializer.ESRIComFromString (strFeature1) ;
    IGeometry pGeo2 = ESRIComSerializer.ESRIComFromString (strFeature2) ;
    pTopo = (ITopologicalOperator) pGeo1;
    pGeoBuffer = pTopo.Difference (pGeo2) ;
    return ESRIComSerializer.ESRIComToString (pGeoBuffer) ;
}
```

Web Services can be used in network applications after being developed. After adding a reference of the above Web Services into the .Net development environment, .Net will automatically generate the proxy class of Web Services, and then the methods provided by Web Services can be used like local methods. At the same time, .Net will automatically generate a WSDL file, which gives the address, method definition, and other related contents of Web Services. Buffer analysis is taken as an example to illustrate the use of Web Services, and the codes are as follows:

```
localhost.GISServices gs = new localhost.GISServices () ;
IPoint point;
//Set point
string strPoint = ESRIComSerializer.ESRIComToString (point) ;
string strBuffer = gs.Buffer (strPoint, 1000) ;
IPolygon polygon=ESRIComSerializer.ESRIComFromString (strBuffer) ;
//Process return results
```

Professional analysis and application based on GIS can be realized by combining GIS Web Services and oil and gas storage engineering expertise. For example, some pipeline risk prediction analysis functions can be implemented based on buffer analysis and overlay analysis. Looking at a leak as an example, according to the severity and location of the leak, a buffer analysis is performed on it, and then the analysis result and the administrative divisions are analyzed for overlapping. In this way, the affected towns, affected persons, and the severity of the impact can be obtained so that corresponding measures can be taken in case of emergency. The schematic diagram is shown in Fig. 5.4.

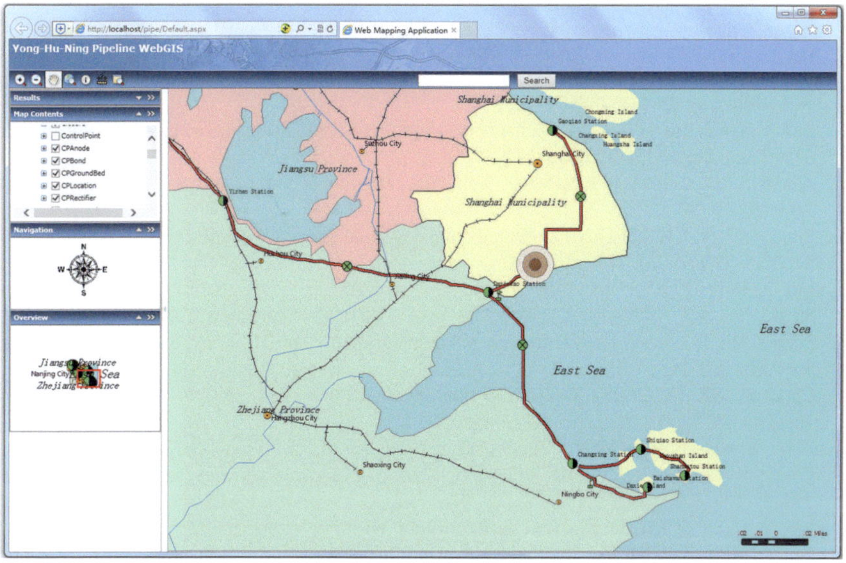

Fig. 5.4 Pipeline leakage risk prediction analysis

References

1. Alesheikh A, Helali H, Behroz H (2002) Web GIS: technologies and its applications. In: Proceedings of the ISPRS technical commission IV symposium 2002
2. Song G, Zhong E (1998) WebGIS-Internet-based geographic information system. J Image Graph 3(3):251–254
3. Yang C, Wang Y (2001) Review of the main technologies of WebGIS. J Image Graph 6(009):886–894
4. Wu L, Liu Y, Zhang J, Ma X, Wei Z, Tian Y (2017) GIS: principles, methods and applications. Science Press
5. Liu N, Liu R (2002) The principle of web GIS and its application. Science Press
6. Wu L, Zhang J (2001) The system and structure of internet-based GIS. Geogr Territ Res 17(4):20–24
7. Zhang C, Sun X (2002) Comparison of models of some current web GIS. Comput Eng Appl 38(15):77–79
8. Zhang X, Ma L, Zhang Q (2005) GIS database. Science Press
9. Wu L (2015) Thinking on sharing 3D GIS data by web service following OGC standard. In: 2nd ISPRS international conference on computer vision in remote sensing, CVRS 2015, April 28, 2015–April 30, 2015, Xiamen, China, 2016. Proceedings of SPIE—the international society for optical engineering. SPIE, p et al; Purdue University; University of Waterloo; Xiamen Municipal Land Resources and Housing Society; Xiamen Municipal Science and Technology Association; Xiamen University. https://doi.org/10.1117/12.2234804
10. Zhao L, Liu S, Li J, Xu H, Luo X (2010) The research of information publishing platform in Web GIS based on Applet/Servlet mode. In: 2010 2nd conference on environmental science and information application technology, ESIAT 2010, 2010. IEEE Computer Society, pp 541–544. https://doi.org/10.1109/esiat.2010.5568876
11. Zhang X, Li G, Lan X (2011) Research on WebGIS performance optimization. In: 7th international conference on wireless communications, networking and mobile computing, WiCOM

2011, September 23, 2011–September 25, 2011, Wuhan, China, 2011. IEEE Computer Society, Communication University of China; Engineering Information Institute; et al; Huazhong University of Science and Technology; Jiangsu University; Wuhan University. https://doi.org/10.1109/wicom.2011.6038689
12. Luo X, Xie Z, Luo J (2013) A mechanism of initiative transmission to send message on WebGIS. Sens Transducers 159(11):236–241
13. Zhang S, Li C (2002) Java-based web GIS research and development. Comput Eng Sci 24(1):54–58
14. Luo Z-Y, Luo J, Lai D-J (2012) Architecture design of plug-in WebGIS using RIA technology. Sci Surv Mapp 37 (6):160–162,110
15. Liu G, Tang D (2009) WebGIS development: ArcGIS server and .NET. Tsinghua University Press
16. Shen J, Wu J, Rong K (2004) Design and application of WebGIS based on WebService. Mod Surv Mapp 27(4):14–16
17. Chen B, Xu M (2003) Web-based computing model: web service. Appl Res Comput 20(1):41–42
18. Zhou W, Mao F, Hu P (2007) The theory and practice of open WebGIS. Science Press
19. Chang Y, Park H (2006) XML web service-based development model for internet GIS applications. Int J Geogr Inf Sci 20(4):371–399
20. Wu G, Liu Z (2006) Design and application of WebGIS based on GIS web service. J North China Inst Water Conserv Hydroelectr Power 27(01):71–73
21. Wu L, Tang D, Liu Y (2003) Interoperable and distributed geographical information systems based on web service. Geogr Geo-Inf Sci 19(4):28–32
22. W3C (2004) Web services architecture requirements. http://www.w3.org/TR/wsa-reqs/. Accessed 07 Oct 2017
23. Shodjai P (2006) Web services and the Microsoft platform. http://msdn.microsoft.com/en-us/library/aa480728.aspx. Accessed May 03 2018
24. Adams H, Gisolfi D, Snell J, Varadan R (2002) Best practices for web services: Part 1, back to the basics. http://www.ibm.com/developerworks/library/ws-best1/index.html?S_TACT=105AGX52&S_CMP=cn-a-ws. Accessed 15 Aug 2017
25. Chai X (2001) Architecture web service: what is a web service? https://www.ibm.com/developerworks/cn/webservices/ws-wsar/part2/
26. Wolter R (2001) XML web services basics. http://msdn.microsoft.com/en-us/library/ms996507.aspx. Accessed 07 Oct 2017
27. W3C (2004) Web services architecture. http://www.w3.org/TR/ws-arch/. Accessed 10 Sep 2017
28. W3C (2004) XML schema. http://www.w3.org/XML/Schema. Accessed 09 Oct 2017

Chapter 6
Summary

The construction of long-distance pipelines is currently leaning toward large-scale, systematic, and networked development. With the continuous expansion of pipeline construction, traditional survey and design and construction management methods can no longer meet the needs of pipeline construction and operation management. The concept of Digital Pipeline is derived from the concept of Digital Earth. The introduction of the concept of Digital Pipeline provides new means, methods, and concepts for the construction and management of long-distance pipelines.

The goal of Digital Pipeline construction is to conduct modern, scientific, and all-digital management of the pipeline design, construction, and operation throughout the life cycle of the pipeline by using high-tech means. It organizes all of the information about each point on the pipelines on the basis of geographic coordinates and then builds the information models for the entire pipelines. Information about any point and any aspect of the pipeline can be obtained quickly, visually, and completely through the model so as to achieve an ideal state that any information is at hand. Digital Pipeline is a platform for providing efficient data collection and processing tools for pipeline feasibility studies, survey and design, and construction and operation management. It is also a support system for digital management and decision-making.

The construction of the Digital Pipeline will be a new revolution in the traditional pipeline construction method and construction concept. The application of the Digital Pipeline technology will contribute to ① optimized route selection in the pipeline survey and design, ② more scientific, quick, and effective construction organization and deployment in pipeline construction, ③ and more efficient and safer pipeline operation management. In a word, Digital Pipeline will greatly improve the level of exploration and design, and then help to realize the scientific and standardized pipeline construction and operation management systems.

The construction of Digital Pipeline, a very broad field, involves a large number of disciplines with knowledge in geographic information systems, remote sensing, global positioning systems, data acquisition and monitoring systems, 3D visualization, virtual reality, spatial data storage, network technology, oil and gas storage and transportation engineering, etc. Since Digital Pipeline is a relatively new concept,

Z. Li, *Pipeline Spatial Data Modeling and Pipeline WebGIS*,
https://doi.org/10.1007/978-3-030-24240-4_6

there are a few applications and practical experiences for reference, and related information is quite limited. Under such circumstances, it is very challenging to carry out research on Digital Pipeline implementation.

With the rapid development of network technology, the author believes that networked Digital Pipeline will be the trend and direction of Digital Pipeline construction. By relying on scientific research and practical engineering projects such as "Yong-Hu-Ning Long-distance Pipeline Digital Project" and "Developing 3D GIS Web Services and Applying It in Oil and Gas Pipeline Geographical Information System", the author carried out a large number of long-term "Web-based Digital Pipeline" research and practice, and obtained plentiful research results. The research results of this book (volume 1 of the Digital Pipeline monograph series) are described below.

(1) Established the Pipeline Spatial Data Model (PSDM)

The pipeline spatial database is the core of the Digital Pipeline system. The key to building the pipeline spatial database is to design a proper pipeline spatial data model. The research on pipeline data models in China is limited and still in its early stages. However, in the world, many pipeline companies have started research on pipeline data models beginning in the 1990s. So far, representative pipeline data models around the world include ISAT, PODS, and APDM. However, there are also many differences in the demand for pipeline data models at home and abroad due to different specific conditions of pipeline industries. Pipeline data models such as ISAT, PODS, and APDM are mostly based on pipelines in North America. Therefore, these models may not fit the situation in China. Based on APDM, the author designed and implemented the Pipeline Spatial Data Model in this book by drawing on the design experience of present main pipeline data models of the pipeline industry, combining the actual demands and circumstances of the author's implementation of the Digital Pipeline projects, and considering the current situation of pipelines in China.

PSDM is designed to be a general and extensible pipeline data model to promote sharing and achieve interoperability of data of the pipeline industry. In order to achieve this goal, object-oriented programming is adopted, and PSDM is designed as an object model of three levels: Abstract classes, Core classes, and Entity classes. Abstract classes define the framework of PSDM and define the common features of a pipe feature class. All classes except abstract classes inherit the attributes, relationships, and behaviors from the abstract class. Core classes define the Centerline features, linear referencing, and other key features of PSDM. Entity classes inherit from the Abstract classes and are used to describe specific pipeline features. PSDM contains most of the data of the pipeline industry. The pipeline companies can then add data that is not defined by PSDM by inheriting the abstract classes of PSDM. Since abstract classes of PSDM define the common behavior of pipeline features, the newly added features can be accessed in a consistent manner, thus achieving the versatility of PSDM. The versatility and extensibility of PSDM can help pipeline companies reduce repetitive data modeling, and facilitate the sharing and interoperability of pipeline spatial data.

PSDM has a modular design of pipeline elements. According to different businesses, roles and functions, pipeline elements are divided into modules such as pipeline Centerline facilities, sites and facilities, measurement and testing, crossing defects and inspection, cathodic protection, and risk and protection. Considering most existing pipeline data models do not support or have only limited support for several important pipeline businesses including fire protection, repair and maintenance, and pipeline automation, corresponding modules are added into the PSDM. New modules can be added and unwanted modules can be removed according to the actual demands of the project. A module can also add pipeline elements or remove unwanted elements. In addition, compared with the existing pipeline data models, many important pipeline-related elements are added to the PSDM modules. For example, Unfavorable Geology of Landslide, Debris Flow, and Collapse, and Pipeline Protections of Retaining Wall, Ecological Protection, Slope Consolidation, etc. are added to the Risk and Protection module of PSDM.

PSDM supports linear referencing and dynamic segmentation technology. With the development of GPS, GIS, and RS technologies, the absolute positioning of pipeline features is increasingly accurate, and the positioning of pipeline features through linear referencing still has unique advantages. PSDM supports both of the positioning methods. Based on linear referencing, PSDM also supports dynamic segmentation of the pipeline features. According to research on the linear referencing method and dynamic segmentation algorithm, the linear referencing methods suitable for the pipeline system include milepost referencing method, 3D projected method, 3D slack chain method, 3D geoid method, and 2D projected method; and the dynamic segmentation algorithm suitable for pipeline system is interpolation method.

The real-time data of the pipeline system is of vital importance to the operation management of the pipeline. Therefore, support for the pipeline real-time data is fully considered in the design of PSDM. In PSDM, the real-time data of the pipeline is divided into two types: (A) Real-time states and readings of certain installed inherent pipeline facilities, equipment, and instruments and apparatus; (B) Real-time readings of temporarily installed instruments and apparatuses measuring instantaneous working conditions of the pipeline operation. These two types of real-time data are designed and processed, respectively. The key problems of the pipeline real-time data acquisition, display, integration, etc., are also addressed.

The third generation spatial data model Geodatabase is taken as the implementation objective of PSDM. In the design process of PSDM, the design concept, data structures, and logic models of Geodatabase are fully made use of so as to obtain the advantages of Geodatabase. There are several ways to implement PSDM as a Geodatabase. The author takes the following strategies to implement PSDM as a Geodatabase: first, model the PSDM object through UML, and then generate the corresponding dataset of the Geodatabase through the ArcGIS Case Tools extension.

(2) Proposed a scheme for the implementation of pipeline WebGIS based on Web Services and Component GIS

The purpose of the pipeline WebGIS is to realize the input, release, query, management, analysis, and output of network-based pipeline information. The traditional methods for implementation of WebGIS mainly include the CGI method, Server API method, Plug-in method, ActiveX method, and Java Applet method. However, these methods have common shortcomings—they are unable to realize the sharing of heterogeneous spatial data, GIS interoperability, cross-platforms, or the sharing of functional resources. To address these problems, this paper proposes a WebGIS implementation method based on Web Services and Component GIS.

Web Services are applications entirely based on Internet standard protocols or specifications (e.g., XML). Client programs can access them through standard protocols such as HTTP and SOAP. Web Services are application logic for providing services and are based on widely accepted standard protocols, so they can be supported by any system or development language. Web Services work at the code level and can be called by other software and exchange data with other software in order to form an application system that can interact with users. Therefore, the GIS based on Web Services can solve problems related to the GIS data integration and share within a wider range and at a higher level.

At the same time, Component GIS has become the mainstream application of GIS development. The basic idea of Component GIS is to divide the major functional modules of GIS into several functional components. Each component performs different functions. These components can be developed by different software providers in various languages, all with the intent of supporting component technology without special requirements for the development environment. GIS components, and other non-GIS components, can be integrated through visual software development tools to form GIS applications in the end.

Based on the analysis and research of the application mode of Web Services and Component GIS, this book proposes a method. GIS functions are published with Web Services as the application framework while Web Services are implemented by Component GIS. Then, GIS functions published by Web Services are used to build the pipeline WebGIS. This method can not only realize the GIS interoperability by using Web Services but also has many advantages of Component GIS including efficient and seamless integration, flexible structure, low development costs, no need for special GIS development language, and high performance and reusability.

Since Web Services can only pass and return basic data types, this book presents a method—COM interfaces and objects are serialized first and then used as Web Services parameters and return values, thus achieving seamless integration of Web Services and ComGIS.

Digital Pipeline is a field with wide Coverage and strong applicability. It involves multiple disciplines, large data volume, and complex technology implementation. Besides the contents of the pipeline data model and the pipeline GIS which were introduced in this book, other aspects of Digital Pipeline include pipeline SCADA, integration of pipeline GIS and pipeline SCADA, and OLE for Process Control and pipeline real-time data acquisition. Extensible 3D and pipeline network virtual reality system will be elaborated in the volume 2 of the "Digital Oil and Gas Pipeline: Research and Practice" monograph series—"Pipeline Real-time Data Integration and Network Virtual Reality System".

Index

Z. Li, *Pipeline Spatial Data Modeling and Pipeline WebGIS*,
https://doi.org/10.1007/978-3-030-24240-4

I

L

M

O

P

R

S

T

MIX
Papier aus verantwortungsvollen Quellen
Paper from responsible sources
FSC® C105338

If you have any concerns about our products,
you can contact us on
ProductSafety@springernature.com

In case Publisher is established outside the EU,
the EU authorized representative is:
Springer Nature Customer Service Center GmbH
Europaplatz 3, 69115 Heidelberg, Germany

Printed by Libri Plureos GmbH
in Hamburg, Germany